Nature's Heartland

IOWA STATE UNIVERSITY PRESS / AMES

Nature's Heartland

Native Plant Communities of the Great Plains Illustrated in Seasonal Color

A Photo-Essay of Woodland Plants and Prairie

Foreword by Charles E. Little

by William Boon and harlen groe

A note on the authors appears on page 360.

Color separations by Pioneer Graphic through CGI (Malaysia) Sdn. Bhd.
Printed and bound in Singapore by Eurasia Press (Offset), Pte. Ltd.

♾ This book is printed on acid-free paper.

First edition, 1990

Library of Congress Cataloging-in-Publication Data

Boon, Bill, 1933–
Nature's heartland : native plant communities of the Great Plains / Bill Boon and harlen d. Groe. – 1st ed.
p. cm.
Includes bibliographical references
ISBN 0-8138-1163-5 (alk. paper)
1. Botany—Great Plains. 2. Plant communities—Great Plains.
I. Groe, harlen d., 1945– II. Title.
QK135.B66 1990
582.0978—dc20 89-26952

Dedication

This book is dedicated to the Ohio Buckeye tree that provided the city of Ames, Iowa, and its visitors and residents with a bountiful display of spring flower color that was matched only by its red-maroon to orange fall display. The tree was removed in the summer of 1984 to make room for commercial development.

Acknowledgements

While the majority of photographs in this book were taken by the authors, we wish to acknowledge the contributions of four colleagues who also provided photographs:

Robert W. Dyas, Distinguished Professor, Landscape Architecture, Iowa State University (several photographs of prairie communities)

Gary L. Hightshoe, Professor, Landscape Architecture, Iowa State University (aerial photographs of plant communities)

Paul F. Anderson, Professor, Landscape Architecture, Land Use Analysis Laboratory, Iowa State University (aerial photograph of windbreak community)

Jean E. Olson, Extension Landscape Architect, Iowa State University (closeup photograph of specimen plants)

Our special appreciation goes to Rhonda Taggart, calligraphic artist, for her artistic talent in designing the lettering for the scientific and common names and the number style.

We are grateful to several persons for their editorial assistance: Barb Brinkman for typing the text and JoAnn Burg for typing the individual botanical descriptions. Special thanks to Tina Herzog, Nancy Santon, and Frank Clark for typing early versions of plant descriptions.

Individual thanks to the people at Iowa State University's Photo Service for their patience with our requests to redo many photographs again and again.

Seasonal photographs of the many individual plants were taken on the Iowa State University campus or within the surrounding community except for the following:

Dawes Arboretum, Newark, Ohio—
American Beech (winter crown)

Botanical Gardens, University of North Carolina, Asheville—
Yellow Birch (winter crown)

Clarence and Alda Hightshoe, Iowa City, Iowa—
Eastern Wahoo (summer, fall fruit, and crown)

Holden Arboretum, Mentor, Ohio—
Black Willow (winter crown)

We wish to express our thanks to all who have assisted and waited patiently for the publication of this book. Our thanks as well to Iowa State University Research Foundation, for making the production and publication of this book possible. We are truly grateful.

Contents

Foreword

Thomas Jefferson believed it would take forty generations to settle the continental vastness of the interior of America. As it turned out, it took scarcely four. Just ninety years elapsed between the explorations of Lewis and Clark commissioned by President Jefferson in 1803 and the presumed "closing" of the frontier, which famed historian Frederick Jackson Turner declared had happened already in his famous speech at Chicago's Columbian Exposition in 1893.

This was a great victory for Manifest Destiny, to be sure. It was also a great victory for the steel moldboard plow. No true progress—in the nineteenth-century sense of the word—was possible until cultivation could be brought to the resistant plains, the matted climax ecosystems of native perennial grasses, tall and short, that dominated the billion-acre heartland of America. These tough prairie sods had thwarted the old-fashioned iron and iron-clad wooden plows based on European designs, confining agriculture to the loose alluvial soils along water courses and the easily plowed hillsides, once they had been cleared of trees.

Then came John Deere, a Vermont blacksmith who in 1839 had invented a way to mass produce a steel plow, whereupon he established a factory in Illinois that could turn them out by the thousands. The Deere plows and their imitators were light, strong, inexpensive, and could, with a stout team, cut the prairie into turves and lay the thick sod on its back at last.

Thus was the great land settled, by steel plow and ax, an acre at a time. The Homestead Act, signed by President Lincoln in 1862, transferred 147 million acres of federal domain to the ownership of 1.6 million families. And suddenly—indeed, in any historical sense, almost instantaneously—the American plains were ours. Manifest Destiny, so fervently sought by a new nation, was at last achieved—not by conquering western mountain vastnesses or scorching deserts, but by conquering grass. The heartland was converted decisively and permanently to a vanquished, grossly simplified *agro*-ecology of fragile annuals—wheat, corn, soybeans. Soon the single plowshares gave way to tractors and chemicals; the yields increased; and the cities grew.

The result was, and is, the most productive agricultural region on the face of the earth, with an accompanying multi-national, big-business economy to match. We must acknowledge that as an achievement. But there has been a cost, too, for at no period in human history has so much land been so utterly transformed ecologically in so short a time. Now that a century has passed since the transformation it would appear that there is virtually no one left who can even remember how it used to be. How the grasses of the tallgrass prairie were tall enough to hide a herd of deer and the panther stalking them; how the shortgrass prairie could survive the pounding thunder of stampeding buffalo; how the lush potholes enticed the breeding waterfowl in sky-darkening flocks each spring; how the ancient hardwood giants, the oak and beech and ash that grew along the banks, shaded the fat pickerel finning lazily in the woodland streams; and how this biological diversity created an indigenous Native American culture that used it and worshipped it at the same time.

But wait. Even if, from the windows of an airliner or a speeding car, the middle-western landscape may look altogether tamed and organized into cornfields and condominiums, unrecognizable to those aboriginal settlers, maybe we had better look again. Indeed, if we can learn to look carefully, perhaps we will find something of great value. And that is the subject of this book: the valuable bits and pieces of the native plant communities that still survive, the grasslands and potholes and woods mercifully skipped over by the rush to plow the land and build the cities. For those who wish to learn of these natural communities, the book you are now holding in your hand is a treasure; it can serve as a kind of Rosetta Stone to our understanding of what we have lost, and of what we can still preserve. Conceived and written by a gifted teacher of the art and science of landscape, Bill Boon, and beautifully designed by harlen Groe, *Nature's Heartland* helps us to imagine how it used to be, this vast primeval ecosystem, by helping us to piece it together in our minds and hearts (and perhaps even our actions) from the remnants that are left to us.

As you study *Nature's Heartland* you will soon come to understand that this no exercise in ecological nostalgia. There is a subtext here, albeit often concealed, that offers its own creative challenge. Let us find a way, the book insists, to protect all the remnants—even recreate them if possible and patch them together again—so that something of the original ecology can be made to function anew at least in certain areas.

Here then is a starting point, a way to uncover the ecological past and equip us to deal intelligently in the future with the natural landscapes of the Middle West and Great Plains. Those professionally involved with natural history research and teaching will find the comprehensiveness and the scientific precision of this book much to their liking. Moreover, *Nature's Heartland* will doubtless become an invaluable reference work for resource conservationists who wish to persuade their fellow citizens and our political leaders that we should try to recreate some of these ecosystems—an enterprise known as "restoration ecology."

But for most of us, the common readers, *Nature's Heartland* is simply a delight to read and to own, to show to one's children and grandchildren, and to pass along to them—together with the remnants, carefully preserved we hope, of the great ecosystem itself.

Charles E. Little
Kensington, Maryland
May 1990

Preface

Most people have a special feeling about plants in the landscape—especially in spring, when the return of plant color brings an end to winter's gloom, and in autumn, when leaves of yellow, orange, and red shine brightly against the blue sky on a sunny day. (The person who has to rake those colorful autumn leaves may not share such romantic feelings.) However, even the most devout worshiper of plants is not without prejudice and may favor one species over another.

For myself, I have chosen the natives. They are the dependable plants that have always been with us. They have adapted to wet and dry conditions, heat and cold, insects and diseases. They are survivors that have proven themselves worthy of being permanent residents of a given region.

How these native plants must rattle their leaves in delight as we plant ornamental after ornamental only to replace these non-native species again and again because of so-called abnormal weather. The abnormal weather, which claims many introduced ornamental plants in a single season, is just part of a cycle to which the native plants have become accustomed.

We are blessed with many delightful native plants that are attractive during several seasons of the year. Some are exciting and colorful, others not so spectacular, but each in its own way has a special seasonal display. It is with these native plants that I hope *Nature's Heartland* inspires you. I hope you will come to love and appreciate native plants as I have.

Today, the native plants of the Midwest are only found along streams, creeks, and rivers and on steeply sloping land. That is because almost all land that is suitable for farming has been denuded of its native timber or prairie plants at one time or another and put to agricultural use. As a result, relatively undisturbed tracts of native prairie and woodland plants are becoming scarce.

Each year much of our remaining natural landscape is further depleted by agriculture, deforestation, and urban expansion. Many of the remaining native plants are growing in unstable and deceptively fragile areas. If these remaining areas are robbed of their vegetation, they will erode without proper care.

The native plants already growing in woodland areas and on open prairie sites are the best plants for those particular soil types and climates. Unfortunately, many of the native plants on such sites are seen as having little value when exactly the reverse is true. In addition to protecting the soil, native plants add beauty by providing variety and contrast to the Midwest's vast tracts of farmland. By providing habitat for wild birds and other animals, these areas provide recreation for the hunter, the amateur biologist, and anyone who loves to walk through natural areas.

The restoration of native vegetation, whether because of private or public concern, is becoming a necessary trend in design. This concern and design approach is found throughout the United States in the restoration of urban and rural landscapes. We, as individuals and as a nation, are losing too much native vegetation to maintain a healthy "web of life." The problem: how does one restore land or landscapes that have been altered? The answer: one needs to understand the natural environment as a system.

A knowledge of plant communities and how they interrelate is necessary in order to emulate them. If land were allowed to remain fallow, it would restore itself through the natural process of succession. But this would take hundreds of years! The restoration process can be greatly hastened by using the knowledge of plant community relationships to make informed selections of native plants.

The remaining areas of native plants would best be left untouched. At the very least, these areas should not be altered for short-term gain, either public or private. Goals for long-range management and use should be established for them. They should be used for wildlife travel corridors and refuges, for soil conservation programs, and for public and private outdoor recreation.

If you have difficulty justifying ownership of timberland or prairie because it doesn't produce anything, then I hope you will look this book over carefully. It will introduce you to plants that flourish on steep slopes, in wetlands, and on other untillable ground. It will encourage you to watch these plants change through the seasons. It might even help convince you to save these threatened and irreplaceable treasures: our native plants.

William C. Boon

Introduction

Nature's Heartland is a book about woodland and prairie plants that are native to the Midwest. The Midwest is an area where plants from several regions intermingle. Many species found growing here are also common to Eastern Deciduous Forest, Northern Lake Forest, Southern Pine Forest, and the Grasslands that reach westward into the Rocky Mountains. People from these regions will find many familiar plants in this book, as the Midwest is a region of transition between different vegetation types.

Nature's Heartland presents plants in their seasonal moods and habitats. The plants are grouped by plant communities. In nature, plants commonly grow in association with a particular group of other compatible plants. These plant associations, or communities, are consistently found where specific environmental conditions (e.g., soil, topography, climate, and water) exist. Grouping plants according to their natural communities helps the amateur find and identify plants in the field. It also aids the professional who is searching for plants for a natural design setting.

Each chapter in *Nature's Heartland* deals with a distinct plant community. Each begins with a written description of the community, a map showing its geographic distribution, and a list of plants found within that community. Photographs give an overview of the community, showing its visual character through aerial views and selected close-ups of individual plants representative of that community. Each species page that follows focuses on a particular member of that plant community, with a general statement about the plant, a botanical description, a species distribution map, and photographs portraying the seasonal moods of that species. Color, texture, and the dramatic change of deciduous plants throughout the four seasons are shown. Other photographs include details of bud and bark, leaf (summer and fall), flower, and fruit. Exceptions to this format are made for evergreens, since they show little seasonal change, and for woodland flowers and prairie plants, which are shown up close.

Photographs selected for this book show plants growing in their natural surroundings wherever possible. A few trees, such as Ohio Buckeye, were photographed in urban, rural, or campus settings. An overwhelming majority of the plants were photographed in Iowa's state and county parks and preserves, including the Ledges and Steamboat Rock state parks; Ames High School, Doolittle, and Black's prairies, and the Boone Railroad Right-of-Way Prairie. Many woodland flowers were photographed at the Ledges and Wapsipinicon state parks and at Pammel Woods on the Iowa State University campus. Some 1478 photographs of 11 plant communities show the visual character of 202 trees, shrubs, vines, woodland flowers, and grasses.

Plant descriptions and botanical terminology have been kept to a minimum in the five descriptive categories (habitat, form, foliage, flower, and fruit.) Individual characteristics or unique plant qualities are in capital letters (emphasized) to encourage an awareness of a plant's visual character, e.g., BURLY fringed acorns (Bur Oak) and SHAGGY bark (Shagbark Hickory).

The glossary in the back of this book, a slight modification from the normal, is in an order to help understand and distinguish the most commonly encountered plant characteristics. Within each of these visual keys we have loosely arranged the plant features in similar arrangements (e.g., leaves: simple, compound; flowers: clustered, solitary). Each visual plant key follows the botanical description that appears with each plant photo essay (habitat, form, foliage, flower, fruit).

Under the key, Form, descriptions of plant form, branching type, and bud and bark types may be found. Similar groupings of plant elements occur for Foliage: leaf type, leaf shape, leaf margin, surface and veination, leaf tip, and base. Flower keys consist of flower type and flower parts, followed by fruit type. Each group of visual keys is accompanied by two or three photo-illustrations that may help illustrate a specific plant character to ensure a quick and accurate identification.

The use of scientific names is commonly preferred over the use of common names by botanists, foresters, horticulturists, landscape architects, and other professionals throughout the world. Synonyms often exist and sometimes reflect differences of opinion among environmental professions and technical books. The first word of a plant's scientific name (such as *Quercus*) is the genus, or the general group, to which the plant belongs (in this case, the oaks). The second word (such as *macrocarpa*) denotes the species and is the specific name (in this case, Bur Oak). Scientific names used in this book agree with those presented in *Standardized Plant Names* (Kelsey and Dayton, 1942).

Geographic distribution maps show where each species is found in its native environment. Most of these maps have been adapted from those published in the *Atlas of United States Trees* (Little Jr., 1971) and *Native Trees, Shrubs and Vines for Urban and Rural America* (Hightshoe, 1988). Distribution ranges for a number of non-native (European) plants, such as evergreen trees of the windbreak community and shrubs of the Oak-Hickory Community, are shown with alternating color bands throughout their ranges.

In nature, more than one major vegetation type can be found within a plant association. An overlapping range is depicted by the Pine-Fir-Birch and the Maple-Linden communities (i.e., the Boreal Forest and the Great Lakes Forest). Vegetation and plant species maps, along with manuals and published studies, reflect the most accurate and recent information available on a species' natural range.

The maps on the preceding pages show the hardiness zones and natural forest vegetation types of the United States and southern Canada.

The diverse landscape character of our natural vegetation is influenced by many factors. The climatic differences of specific geographic areas (e.g., floodplains and mountains), environmental conditions (e.g., length of day and sun angle), and changes in altitude are all factors in determining plant hardiness zones, as are differences of temperature, soil, and moisture.

These climatic variations or zones are averages, and exceptions should be noted. Altitude differences may lower the temperature by three degrees per 1000 feet locally. In addition, inclines greater than 10 percent may raise or lower zones, in which case the north-facing (cooler) slope is one zone lower and the south-facing (warmer) slope is one zone higher. Climatic variations of big lakes and water shorelines may have an opposite effect.

Whether a plant, given suitable climatic and environmental conditions, will survive and grow is contingent on its winter hardiness (the lowest temperature it can survive). For example, zone 5a includes trees and shrubs that will survive temperatures of zone 5a or the zone south of it. The same plants may also be found in the next zone north (4b), where the environmental conditions or microclimate are suitable.

Vegetation Zones

UNITED STATES DEPARTMENT OF AGRICULTURE:
HARDINESS ZONES

Plant zones are delineated in increments of (10) degrees which show the average temperatures of the United States and Canada. Each zone is further divided into an upper and lower (a and b) area with the lower temperature occurring in the (a) zone.

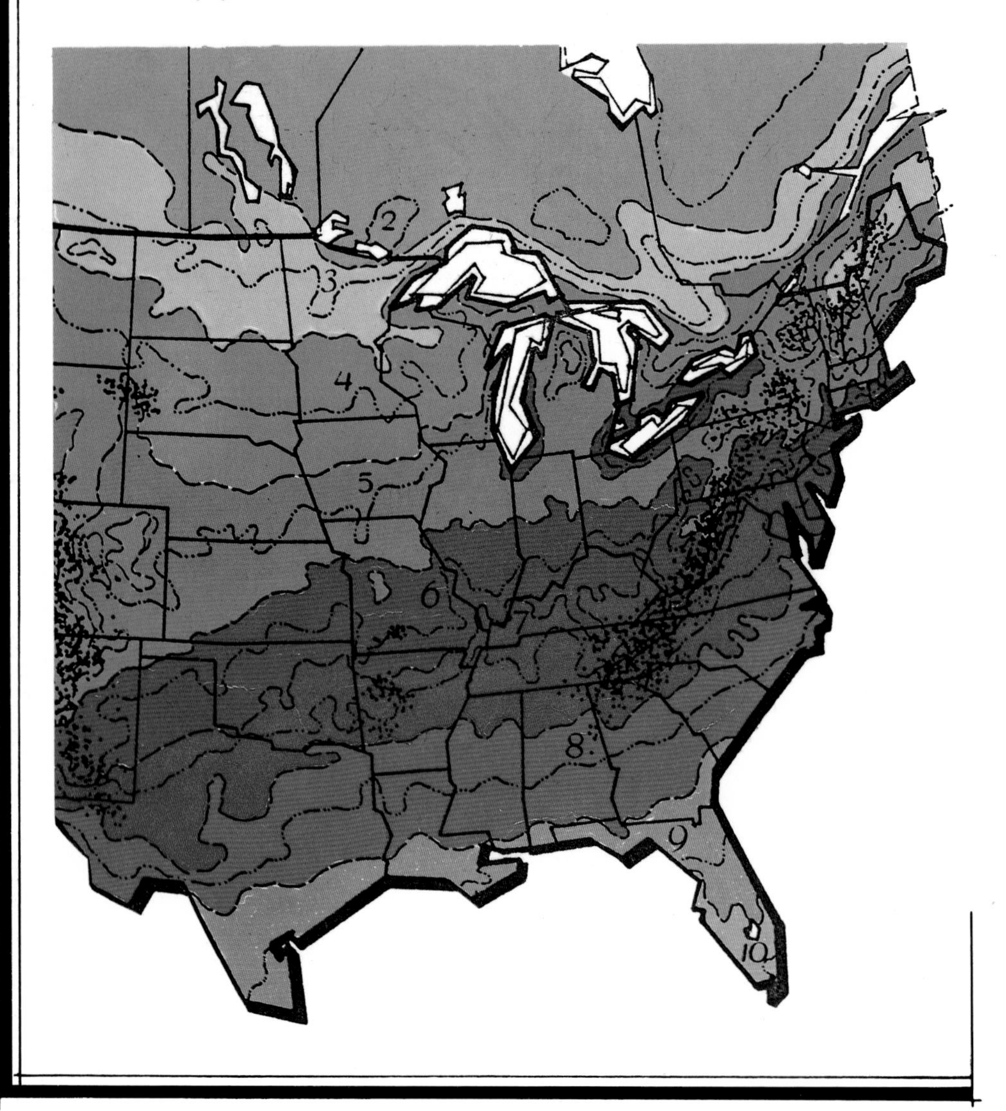

Zone	temperature (approx.)
1	. . . . . . -50°
2a	. . . . . . -50° to -45°
b	. . . . . . -45° to -40°
3a	. . . . . . -40° to -35°
b	. . . . . . -35° to -30°
4a	. . . . . . -30° to -25°
b	. . . . . . -25° to -20°
5a	. . . . . . -20° to -15°
b	. . . . . . -15° to -10°
6a	. . . . . . -10° to - 5°
b	. . . . . . - 5° to 0°
7a	. . . . . . 0° to 5°
b	. . . . . . 5° to 10°
8a	. . . . . . 10° to 15°
b	. . . . . . 15° to 20°
9a	. . . . . . 20° to 25°
b	. . . . . . 25° to 30°
10a	. . . . . . 30° to 35°
b	. . . . . . 35° to 40°

Vegetation Types

Mixed Prairie
Dry (Xeric)
Marsh-Pothole
Wet (Mesic)

Oak Savanna

Oak-Hickory

Maple-Linden

Pine-Fir-Birch
Boreal Forest

Mixed Forest
Maple-Linden
Boreal Forest

Mixed Floodplain
River-Lake Margin

'conterminous' excluding
Hawaii and Alaska

POTENTIAL NATURAL VEGETATION OF THE CONTERMINOUS UNITED STATES (Küchler, '64) defines potential natural vegetation as:
"the vegetation that would exist today if man were removed from the scene and if the resulting plant succession were telescoped into a single moment".

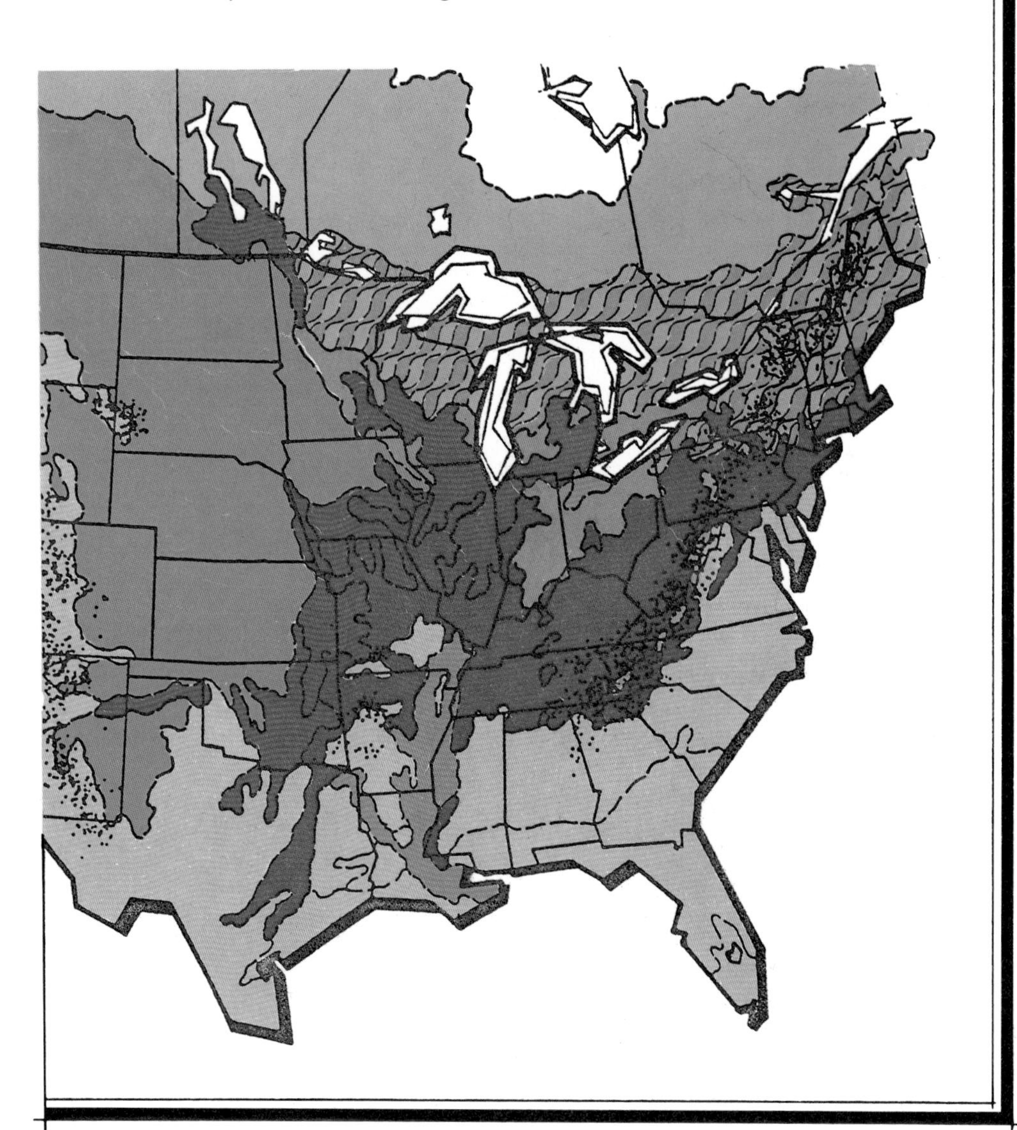

Native Plant Communities

The Dry Prairie, or shortgrass prairie, is a community where water and soils are the primary limiting factors. The shortgrass prairie is primarily found on the western plains of eastern Colorado and western Nebraska, but there are also remnants mixed in with the tallgrass prairie in Iowa, Wisconsin, and Illinois. These prairies are typically found on rocky knolls and dry sloping woodland clearings where the soil is shallow or sandy and moisture is lacking.

In these shortgrass prairies, plants have adapted to the extreme heat and a limited supply of water. Some grasses fold or roll their leaves during times of moisture stress, due to increased osmotic pressure. The majority of plants are perennials that develop efficient root systems as they mature. Roots of the various species absorb water and nutrients at different levels. While some forbs and grasses limit their root systems to the first foot of soil, others extend 17 feet or more below the surface in search of moisture. Some surface areas seem to be relatively uncrowded, but the network of underground parts is so dense that invading plants are unable to compete.

Due to extreme competition, even perennials fruit sparingly or not at all. They often take several seasons to attain the amount of growth that is possible in a single season where competition is less intense.

Dry Prairie Community

Grasses:

Shrubs: (less than 6′)

Forbs:

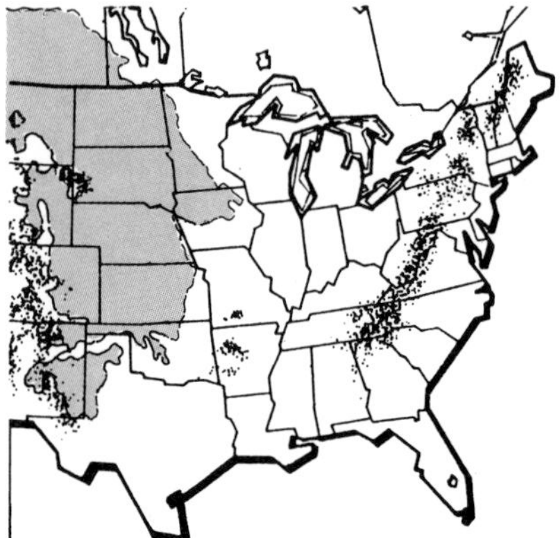

Little Bluestem is a bunchgrass that is common in all states but four of the Pacific Northwest. And it exceeds in abundance all other upland grass species combined (Phillips Petroleum Company, 1963). It is considered a medium-tall grass with seed stalks only 2 to 4 feet high, but its dense root system may reach 8 feet in depth.

Little Bluestem has a beautiful reddish fall color and seeds that look like tufts of feathers tucked here and there. Its name comes from the flat shoot at the base of the plant, which has a bluish color in late spring or early summer.

habitat . . .
PRAIRIE. upland mesic, upland mesic-dry, open woodlands, rocky bluffs and sandy soils
•zone – 2b

form . . .
UPRIGHT. erect clumps with tufted yellowish tan stems (2–4′)
•color (spring) bluish stem

foliage . . .
BLADE. narrow, grass-like, lanceolate tan-bronze leaves (10″) with sheath at stem base; slightly folded
•color (fall) reddish tan to copper, PERSISTING through winter
•season – perennial (July–Oct.)

flower . . .
SPIKE. singular cluster (2–2½″), small, feathery (⅜″) with white, bristly hairs; terminal clusters

fruit . . .
GRAIN. small, bluish red (¼″), FUZZY seed head with silky hairs; top (8″) only
•season – maturing early autumn

DRY PRIAIRE

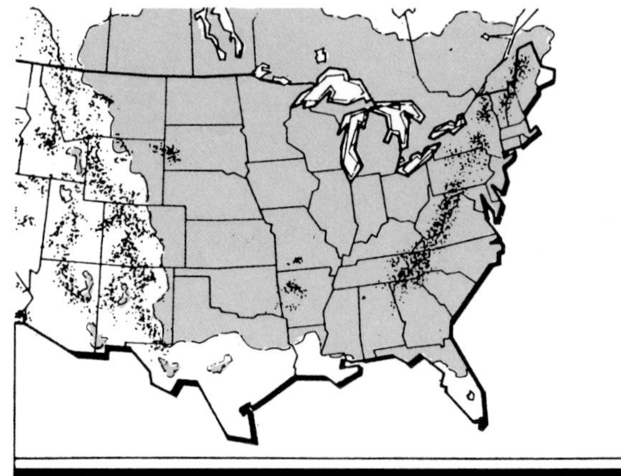

Sideoats Grama is a medium-tall grass that grows 1 to 3 feet. It is widely scattered throughout the prairie and grows in a variety of sites, although it is rarely found in great abundance. When prairies are damaged by extreme drought or by grazing, this grass tends to increase rapidly, even though cattle find it highly palatable (Phillips Petroleum Company, 1963).

Sideoats Grama gets its name from its small seeds, which look like oat seeds, that tend to line one side of the stem.

habitat . . .
PRAIRIE. upland mesic-dry, upland dry and steep rocky slopes
•zone—2b

form . . .
UPRIGHT. erect, wiry clumps (1–2½′)

foliage . . .
BLADE. grass-like curly mass of green leaf blades (12–18″) with bumps along blade edge; hairy
•color (fall) TAN; whitish brown when dry
•season—perennial (June–Oct.)

flower . . .
INFLORESCENCE. small, dark purplish red (5″) spikes with florets (⅜″) fine-toothed along a flat rachis; orange anthers

fruit . . .
GRAIN. tan, OAT-like (⅛″) seed, perpendicular to stem; uniformly LINING one side of rachis
•season—maturing early autumn

DRY PRAIRIE

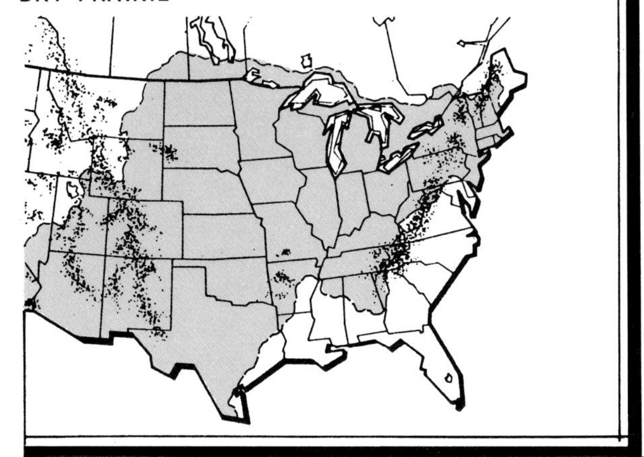

Bouteloua curtipendula · Sideoats Grama

Hairy Grama is a shortgrass that grows 10 to 18 inches tall on shallow, sandy, or rocky soils. It is common throughout the Midwest, but it is never found in pure stands. Its name comes from the fact that both leaves and seed heads are hairy. The seedheads are curved, and their tips are pointed.

Hairy Grama is one of the most nutritious of the grama grasses and is readily eaten by livestock (Phillips Petroleum Company, 1963).

habitat . . .
PRAIRIE. upland mesic and upland mesic-dry
- zone—3b

form . . .
UPRIGHT. dense clumps of erect stems (6–8″)

foliage . . .
BLADE. thin, grass-like, green, hairy leaf blade (10″) mostly curled at base
- color (fall) brown
- season—perennial (July–Sept.)

flower . . .
INFLORESCENCE. small, red-brown (2–4″) spikes with florets (¼″) fine-toothed; rachis extending beyond spike

fruit . . .
GRAIN. hairy, tan, OATS-like (1/16″) seed with ROOSTER-like combs; hairs curved with needle-like point
- season—maturing late autumn

DRY PRAIRIE

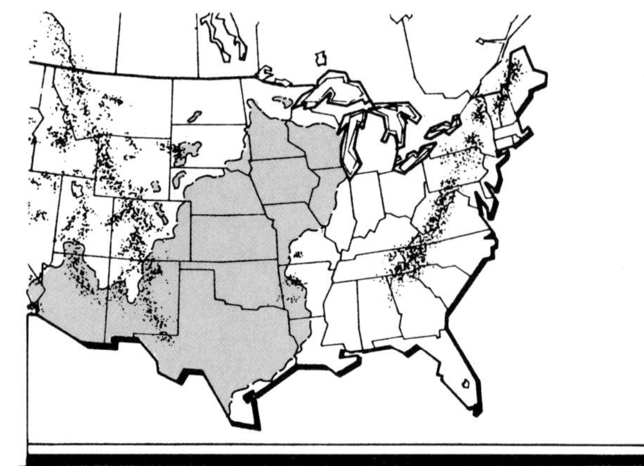

Bouteloua hirsuta · Hairy Grama

Prairie Dropseed grows on hilltops and dry slopes, where it is often interspersed with Little Bluestem and Porcupinegrass. It is a bunchgrass with leaves that curve gracefully. The delicate seed heads appear in midsummer, and the seeds ripen and drop early.

Prairie Dropseed grows about as tall as Little Bluestem (2 to 8 feet), but it tends to flatten out and retain its leaves throughout the winter.

habitat . . .
PRAIRIE. upland mesic-dry, upland dry, open woods edge and sandy soils
•zone – 3b

form . . .
ERECT. slender clumps forming large tufts (2–3′)

foliage . . .
BLADE. slender, grass-like, green leaf blades (8–12″) with basal leaves elongated; arching
•color (fall) tan-BRONZE
•season – perennial (July–Oct.)

flower . . .
INFLORESCENCE. open, many branched (1/4″) florets with short stalk spikelets; terminal

fruit . . .
GRAIN. tan-brown, fine textured; very small
•season – splitting open at maturity

DRY PRAIRIE

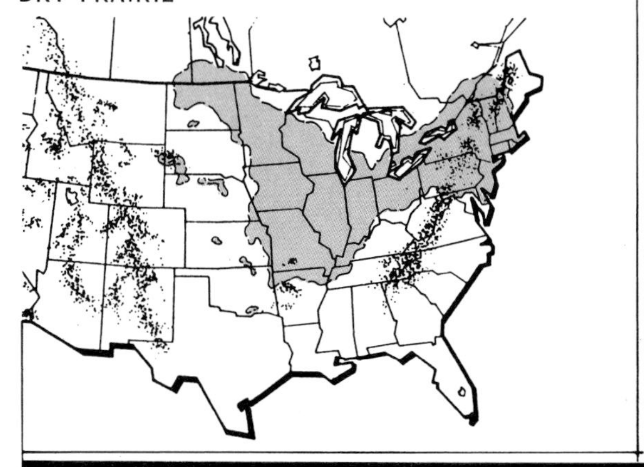

Sporobolus heterolepis · Prairie Dropseed

Porcupinegrass is a bunchgrass that typically grows on dry hilltop prairies. In the spring, its growth is very rapid; it may reach a height of 2 feet by the first of June. Flower stalks follow, and seeds are usually ripe by the second week in June. Unlike the seeds of most other grasses, Porcupinegrass seeds germinate best when planted deep.

The spirally-twisted first segment of the seed arm coils when dry and uncoils when damp. These movements allow the arm to act like a ratchet, screwing the seed into the ground (Pohl, 1978).

habitat . . .
PRAIRIE, SAVANNA. upland mesic-dry, upland dry, open woodlands and rocky sandy soils
•zone—3a

form . . .
ERECT. smooth, upright to arching clumps of stout stems (culm) (2–4′)

foliage . . .
BLADE. narrow, grass-like, long lanceolate green leaves (4–10″) with edges rolled toward upper surface; scabrous above
•color (fall) SILVERY white
•season—perennial (June–July)

flower . . .
INFLORESCENCE. open, grayish green (1½″), NODDING florets, tapering to long slender point; flower head (spikelets) nodding to one side

fruit . . .
GRAIN. small, tan-brown (1/16″), seed-like with sharp-pointed (4–8″) TWISTED awn; self-penetrating
•season—maturing early autumn

DRY PRAIRIE

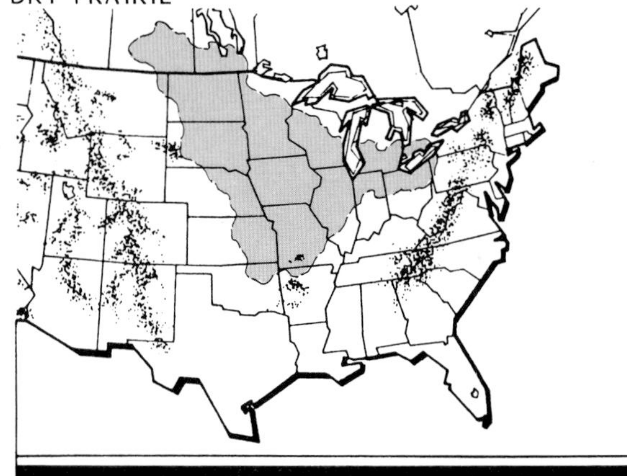

Stipa spartea · Porcupinegrass

Leadplant Amorpha is a small, woody plant of the upland prairies and open woodlands. It is a legume with small, purple flowers that have only one petal each; grouped together, the many flowers along the spike make a colorful display. The leaves are pinnately compound, having many leaflets arranged like feathers along the stem. They are covered with short dense hairs that give the plant a grayish appearance.

Competition with grasses for water is reduced because Leadplant's root system extends 6 to 16 feet in depth (Weaver, 1968). The tough roots spread near the surface and when plowed break with a characteristic snap.

habitat . . .
PRAIRIE, SAVANNA. upland mesic-dry, open woodlands, ridges, rocky bluffs and sandy soils
- zone – 3a

form . . .
GLOBULAR. stems hairy, whitish gray, dense spreading clumps (1–3′)

foliage . . .
ALTERNATE. pinnately COMPOUND, leaflets (15–45), ovate-lanceolate GRAYISH white leaves (4–7″) with entire margin
- color (fall) GRAYISH white
- season – perennial, early summer (June–Aug.)

flower . . .
SPIKE. dense, purple–BLUISH gray (2–3″) clustered with bright orange stamens; terminal

fruit . . .
POD. small, tan-brown, PEA-like husk; tomentose, not opening
- season – maturing early autumn first year

DRY PRAIRIE

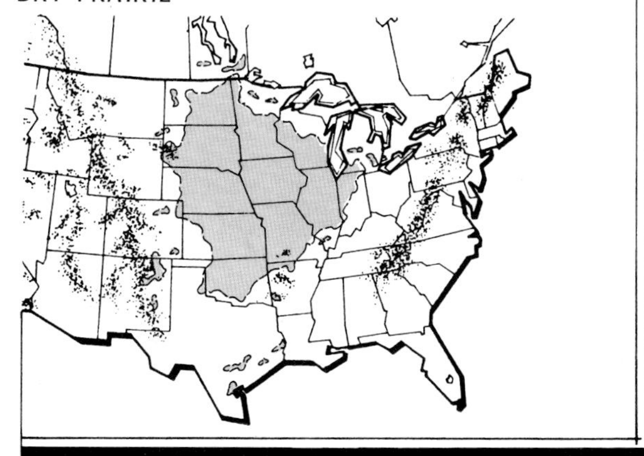

Amorpha canescens · Leadplant Amorpha

Candle Anemone, or Longheaded Thimbleweed, is a perennial flower of dry prairies and open woods. It is a late-blooming anemone, often not flowering until July. The flower is greenish white, and its elongated center resembles a thimble. The thimble later matures to a cottony tuft after frost.

The leaves of Candle Anemone are deeply cut, oak-like clusters arranged in a whorl halfway up the stem. In autumn, the leaves and stems turn a dull red.

habitat . . .
PRAIRIE. dry, open woods, exposed ridges, steep rocky slopes and meadows
•zone – 3a

form . . .
ERECT. slender, gray downy (WOOLLY) surface, multiple stem (1–3′)

foliage . . .
WHORLED. palmate, deeply LOBED gray-green leaves (3–4″) with lobes sharp-pointed, straight margin; stalked basal leaves in whorls (2–3) midway on stem
•color (fall) reddish MAROON
•season – perennial (June–Aug.)

flower . . .
SINGULAR. sepals (5) greenish white (¾″) dia. arising on long (multiple) stalks; without leaves

fruit . . .
CONE. long, green-grayish brown (1–1½″) THIMBLE-like cluster with seed-like achenes; woolly
•season – maturing late summer

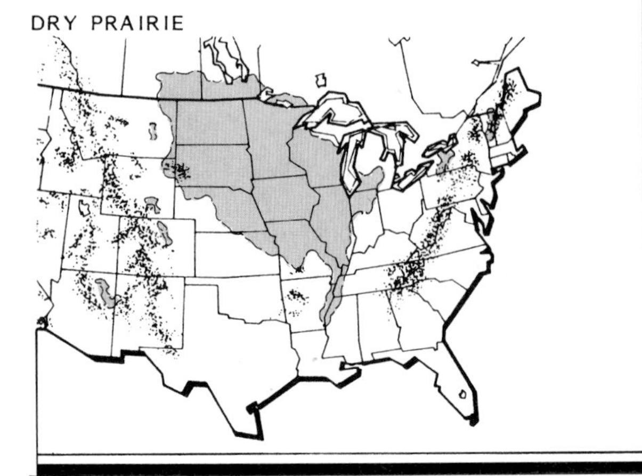

Anemone cylindrica · Candle Anemone

Louisiana Sage, or Silver Sage, is a perennial flower of dry prairies and pastures. It usually grows in patches on poor or shallow soils. The small, yellowish flowers are secondary to the silver color of the entire plant.

The leaves are covered on both sides with a dense coat of hairs that can be removed by rubbing. The reflective silver color and fuzzy leaves are devices the plant uses to help reduce moisture loss in the dry prairie environment.

habitat . . .
PRAIRIE. upland dry, open fields and disturbed soils
•zone – 3a

form . . .
ERECT. ascending, stiff, SILVERY white stemmed herb (1½ to 3′)

foliage . . .
ALTERNATE. linear, lanceolate grayish green leaves (1–2½″) with parted or toothed margin; velvety FELT-like both surfaces, short stalked
•color (fall) GRAYISH green
•season – perennial (July–Oct.)

flower . . .
SPIKE. small, yellowish green (¼–⅜″), BELL-shaped without rays; axillary cluster

fruit . . .
ACHENE. small, light yellow (⅛″), numerous
•season – maturing early autumn

DRY PRAIRIE

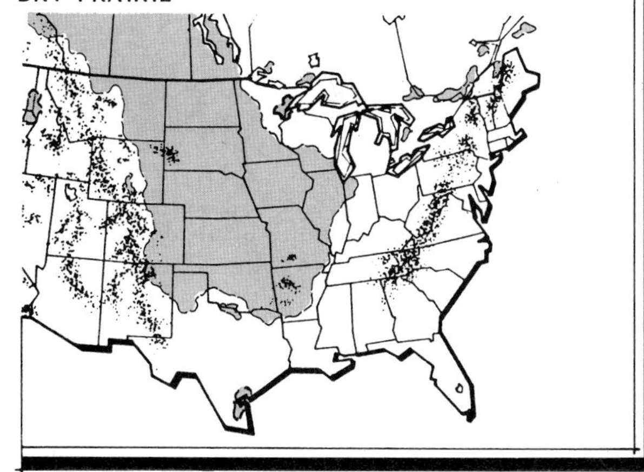

Artemisia ludoviciana · Louisiana Sage

Common Milkweed, an annual, is more widely distributed than any other species of milkweed. Its habitat varies from scrubby areas at the edges of fields and pastures to moist areas of the open grasslands. Typical of all milkweeds, the soft seedpods harden and split open, allowing the white woolly seeds to be transported by the wind.

When the stems and leaves of Common Milkweed are broken, they ooze a milky juice that has a bitter taste. Monarch butterfly larvae feed on this plant and acquire the same disagreeable taste, which they use as a defense mechanism.

habitat . . .
PRAIRIE. upland dry, open meadows, roadsides and disturbed soils
- zone – 2b

form . . .
ERECT. stout, upright, single stemmed herb with MILKY substance (3–6′)

foliage . . .
OPPOSITE. simple, broad-oblanceolate, thick, light green leaves (4–10″) with entire margin; DOWNY gray beneath, exuding a MILKY juice when bruised
- color (fall) gray-BROWN
- season – perennial (June–Aug.)

flower . . .
UMBEL. dull crimson-PURPLE (½″) dense, fragrant (2″) drooping clusters mostly in leaf axils; petals pointing downward

fruit . . .
POD. spiny, grayish green (2–4″) warty (follicle) filled with OVERLAPPING seeds; seeds with tuft of SILKY hairs
- season – splitting open at maturity

DRY PRAIRIE

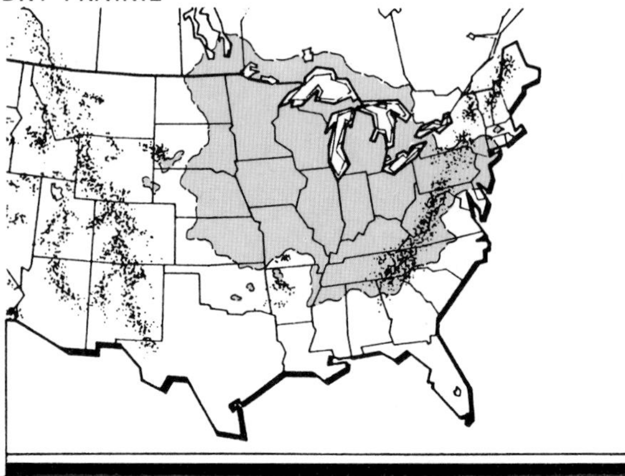

Asclepias syriaca · Common Milkweed

Whorled Milkweed is a common, white-flowered perennial of dry prairies, roadsides, and disturbed sites. It often grows in patches on poor or shallow soils. It is not a welcome guest in pastures, because it is considered poisonous to sheep.

As the name implies, the leaves are whorled along the stem in groups of three to six at a node. It blooms from July to September. The greenish white flowers are clustered at the top of the stems. These clusters become upright pods that split and curl open to release seeds for the wind to disperse.

habitat . . .
PRAIRIE. upland mesic-dry, open meadows, rocky slopes and roadsides
•zone—2b

form . . .
ERECT. slender, simple stemmed, unbranched herb (1–3′)

foliage . . .
OPPOSITE. linear, narrow (3–6) WHORLED, bright green leaves (2–5″) with rolled under leaf margin; sessile
•color (fall) GRAY-brown
•season—perennial (July–Sept.)

flower . . .
UMBEL. small, greenish white, (3/8″) dia., open, FLAT-TOPPED clusters in leaf axils; occurring on upper stem

fruit . . .
POD. slender, green-brown (2–3″) smooth (follicle) long pointed; ERECT with upright stalk
•season—opening at maturity

DRY PRAIRIE

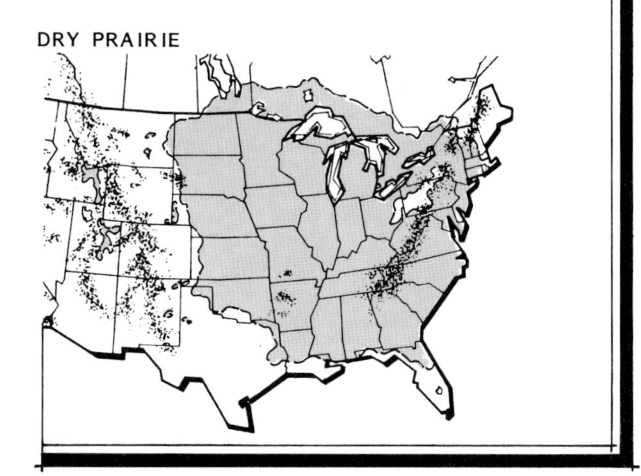

Asclepias verticillata · Whorled Milkweed

Heath Aster is a perennial flower of dry prairies and other open places that blooms from late summer to fall. It is a common aster that is frequently found growing in patches. The bushy, grayish plants bloom in great abundance with miniature, white flowers that look like daisies. Its leaves are also tiny and resemble the leaves of the Heath Plant.

habitat . . .
PRAIRIE. upland dry, open meadows, woods edge and roadsides
- zone – 3a

form . . .
ERECT. slender, open branched, smooth stem with numerous upper leaves (1–3′)

foliage . . .
ALTERNATE. narrow, linear-lanceolate, HEATH-like yellow-green leaves (1–3″) sharp-pointed with entire margin; velvety, often CLASPING stem
- color (fall) yellow-green
- season – perennial (Aug.–Oct.)

flower . . .
DISK. small, yellow, (3/8″) dia. with narrow (15–25) white rays; flower heads densely clustered

fruit . . .
ACHENE. small, tan-brown, dry seed
- season – maturing late autumn

DRY PRAIRIE

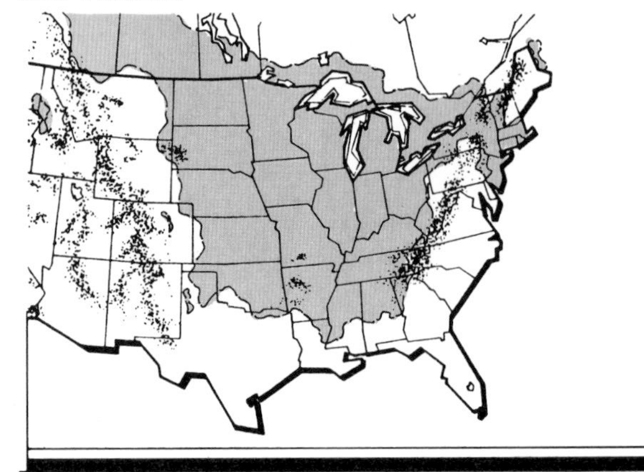

Aster ericoides · Heath Aster

Smooth Aster is a common perennial flower of the upland prairies, open woods, and roadsides. It is one of the most attractive asters, with blue flowers. Its flowers are a pale blue with a yellow center, and each plant has numerous blooms that appear from August to October.

Smooth Aster gets its name from its smooth leaves that clasp the stem.

habitat . . .
PRAIRIE. upland mesic-dry, upland dry, open meadows and roadsides
- zone – 3a

form . . .
UPRIGHT. stiffly erect to flopping, stout, greenish brown stem (2–4′)

foliage . . .
ALTERNATE. slender, thick, ovate-lanceolate dark green leaves (1–4″) with sharp-toothed, long-pointed margin; rough under surface, heart-shaped base CLASPING stem
- color (fall) gray-BROWN
- season – perennial (Aug.–Oct.)

flower . . .
DISK. small, yellow (3/8″) with narrow (15–30) violet BLUE rays with green tipped bracts; flower heads open-clustered

fruit . . .
ACHENE. smooth, one-seeded with red bristles
- season – maturing early autumn

DRY PRAIRIE

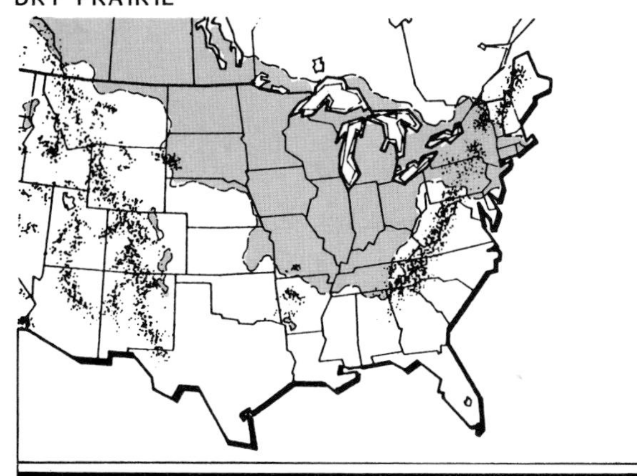

Aster laevis · Smooth Aster

Silky Aster is a perennial flower of dry open prairies. The blue-violet flowers bloom in September and October. The plant has a delicate appearance because of its small, pointed leaves and stiff, wiry, brownish-red stems. The lower leaves fall off during the growing season, leaving only the upper leaves at the time of flowering. The name, Silky Aster, describes the silky hairs on both surfaces of the leaves, which give the plant a somewhat silvery appearance.

habitat . . .
PRAIRIE. upland mesic-dry, upland dry and open wooded banks
- zone – 3b

form . . .
UPRIGHT. delicate, grayish green, wiry, ZIG-ZAG stems (1–2′)

foliage . . .
ALTERNATE. ascending, oblong-elliptic SILVERY green leaves (1–1½″) with silvery white hairs and entire margin; sessile, slightly clasping stem
- color (fall) GRAY-brown
- season – perennial (Sept.–Oct.)

flower . . .
DISK. yellow, (⅜″) dia., compact with spreading floral bracts, rays (½″), lavender-rose, silky; solitary

fruit . . .
ACHENE. small, tan-brown (1/16″), smooth, seed-like, ribbed
- season – maturing late autumn

DRY PRAIRIE

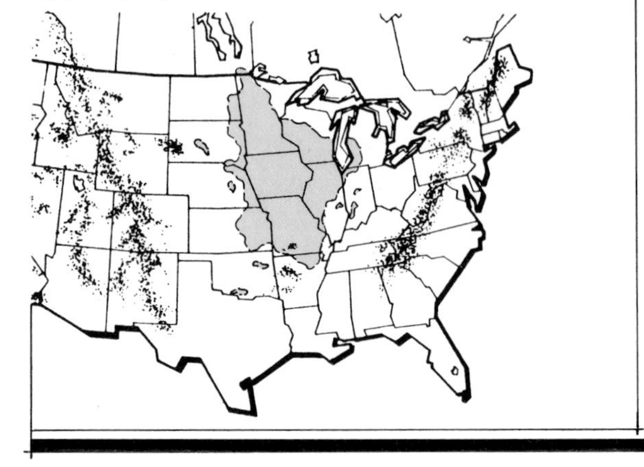

Aster sericeus · Silky Aster

Atlantic Wildindigo is a bushy, perennial legume of dry prairies and open wooded meadows. The wide-spreading branches of this plant are in bloom from June to July. The white or cream-colored, pea-like flowers are held in erect clusters.

The leaves are divided into three large oval-shaped leaflets. Throughout the winter, clusters of large, black-beaked seedpods often remain attached to the naked stems.

habitat . . .
PRAIRIE. upland mesic-dry, upland dry and open wooded meadows
- zone – 4a

form . . .
OPEN. erect, bushy, whitish blue stems with horizontal branchlets (2–5′)

foliage . . .
ALTERNATE. simple, PALMATE leaflets (3), obovate grayish green leaves (1–1½″) with entire margin; smooth, short stalked
- color (fall) bluish gray to BLACK
- season – perennial, herb (June–July)

flower . . .
SPIKE. white (1″) with purplish tinge, erect, PEA-like (6–12″) clusters on angled stems; terminal, lateral (racemes)

fruit . . .
POD. oblong, black (1–2″), beaked, BEAN-like pod; pendent
- season – maturing late summer

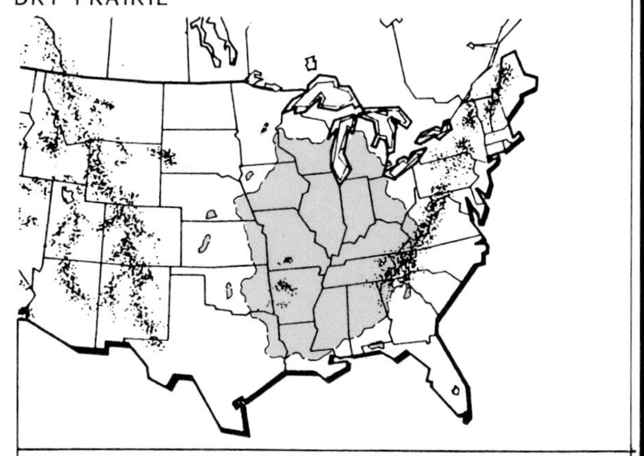

Baptisia leucantha · Atlantic Wildindigo

Plains Wildindigo is a bushy perennial legume that grows in upland prairies on any good, well-drained soil. It blooms in May and June. The dense, drooping racemes of yellow pea-like flowers cause the branches to arch to the ground. The leaves are held in groups of five at a node, with three on one side of the stem and two on the other. These hairy leaves lose their green color by late summer and change to a dark gray or black. The leaves of Plains Wildindigo were used by Indians as a stimulant and for treating cuts and wounds. "A concoction made from the roots was also used as a remedy for scarlet and typhoid fever" (Owensby, 1980).

habitat . . .
PRAIRIE. upland mesic-dry, upland dry and open wooded meadows
- zone – 4a

form . . .
GLOBULAR. low, erect, velvety stemmed with spreading to ascending branchlets (1½ to 3′)

foliage . . .
ALTERNATE. simple, PALMATE leaflets (3), oblanceolate grayish green leaves (1½–3″) with entire margin; whitish hairs both surfaces with (2) large leaf bracts at leaf axils; short stalked
- color (fall) tan-BROWN to black
- season – perennial, herb (June–July)

flower . . .
SPIKE. creamy yellow (1″), PEA-like, drooping to ascending (6–8″) clusters; terminal (racemes)

fruit . . .
POD. oblong, tan-brown (1–3″) long, pointed, PEA-like pod clusters; drooping
- season – maturing late summer

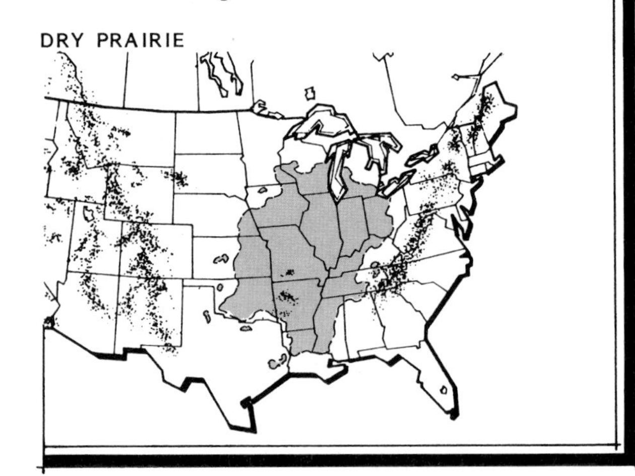

Baptisia leucophaea · Plains Wildindigo

Showy Partridgepea is an annual legume that grows in dry prairies, on disturbed sites, and at the edges of woods. In southern states, it is considered an important honey plant, with nectar obtained not from the yellow flowers but from nectaries at the base of the leaf stalk (Phillips Petroleum Company, 1963).

The leaflets are arranged in pairs along the midrib of the compound leaf. They collapse when touched, giving rise to another common name, Sensitive Plant. When ripe, the pods of Showy Partridgepea split open with an explosive force, throwing the seed several feet.

habitat . . .
PRAIRIE. upland mesic-dry, upland dry, woodlands, open meadows and disturbed soils
•zone – 2b

form . . .
ERECT. slender, zig-zag, prostrate glabrous stems (1–2½′)

foliage . . .
ALTERNATE. pinnately COMPOUND, leaflets (16–30), oblong, bright green leaves (½–¾″) with tiny, bristle tip, entire margin; conspicuous GLAND at leaf base; leaves closing at night and when brushed
•color (fall) yellow-GREEN
•season – annual, herb (July–Aug.)

flower . . .
SOLITARY. large, showy yellow (¾–1″) with (5) unequal petals in leaf axils; dark drooping anthers

fruit . . .
POD. flat, brown (½″), often TWISTING, opening with many flat, brown seeds
•season – maturing autumn

DRY PRAIRIE

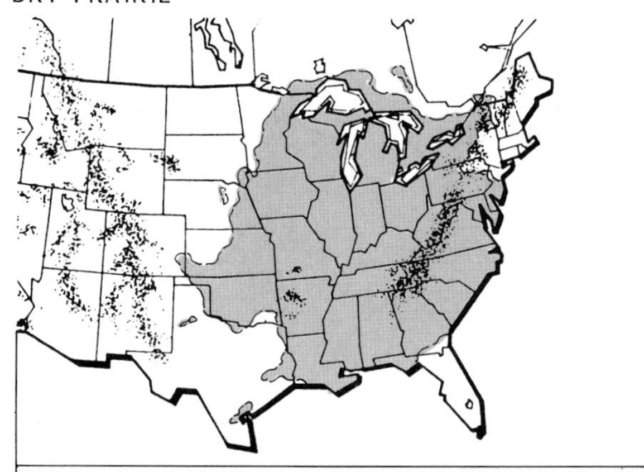

Cassia fasciculata · Showy Partridgepea

Pale Echinacea, or Pale Purple Coneflower, is a perennial of upland dry prairies and wooded slopes. It blooms from June to October. The flowerheads are quite attractive, having delicate purple rays drooping from a large, spiny, cone-shaped center. These spiny seedheads turn brown in late summer and remain standing throughout the winter. The leaves are attached to the plant near its base and are covered with coarse hairs.

Echinaceas are becoming more and more popular as ornamentals and are found in many flower gardens.

habitat . . .
PRAIRIE. upland mesic-dry and open wooded slopes
- zone – 4b

form . . .
UPRIGHT. stout, singular with slightly hairy stems (2–4′)

foliage . . .
ALTERNATE. simple, narrow, lanceolate bright green leaves (4–8″) with entire tapering base; long stalked lower, parallel veined, hairy
- color (fall) green-BROWN
- season – perennial, herb (June–Oct.)

flower . . .
DISK. showy, pink-reddish purple, (1″) dia., compact DOME-like cluster with lavender rays (2–3″) SWEPT back, fading with age; NOTCHED tip

fruit . . .
CONE. spiny, dark brown
- season – maturing late summer

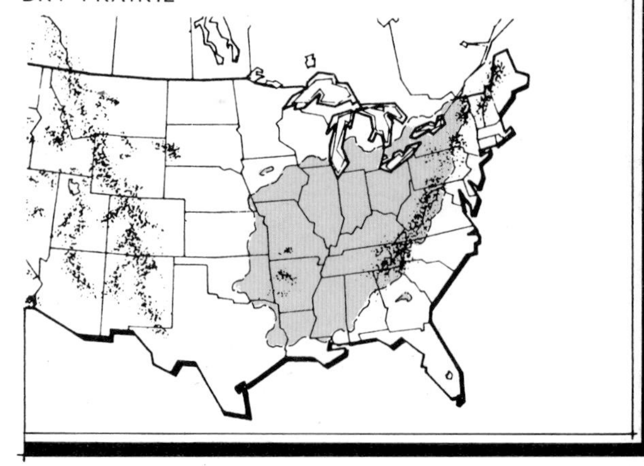

Echinacea pallida · Pale Echinacea

Downy Gentian is a plant of upland prairies, occupying well-drained, sometimes rocky soils. This beautiful little plant blooms in late summer; at times, it even blooms after the first frost. Toward evening and on cloudy days, the flowers of Downy Gentian close as if they had been twisted between the thumb and index finger.

"The gentians were named for Gentius, king of Illyria in 180 B.C. He was given credit for discovering the plants' medicinal values. However, the Egyptians had used the plant several thousand years earlier to treat bites and stings" (Owensby, 1980).

habitat . . .
PRAIRIE. upland mesic-dry, upland dry, open woodland, meadows and roadsides
•zone–3a

form . . .
ERECT. slender, upright, slightly downy, solitary stemmed (6–18″)

foliage . . .
OPPOSITE. simple, oblong-lanceolate light green leaves (1–2½″) with entire margin; stalkless
•color (fall) green-BROWN
•season–perennial, herb (Aug.–Oct.)

flower . . .
CLUSTER. open, purplish BLUE (1–1½″), FUNNEL-shaped with slightly spreading lobes; sometimes twisted tight, clustered in leaf axils

fruit . . .
CAPSULE. tiny, tan-brown seed
•season–maturing early autumn

DRY PRAIRIE

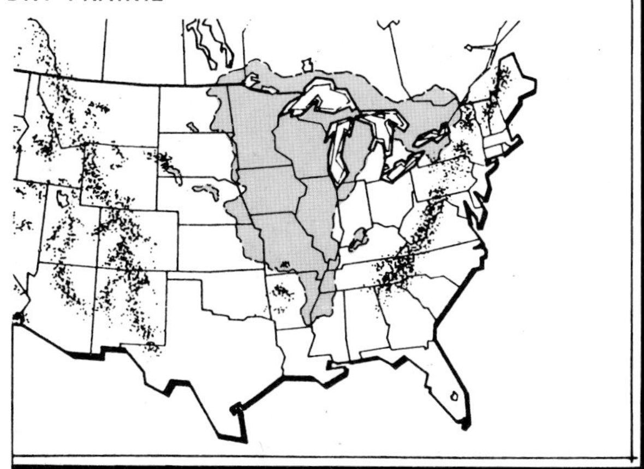

Gentiana puberula · Downy Gentian

Roundheaded Lespedeza, or Bush Clover, is a deep-rooted perennial legume that grows in prairies and open woods throughout the Midwest. The long seedhead stands 2 to 4 feet tall and remains erect even through the winter months. The leaves are clover-like, and in groups of three. The flowers, which are greenish-white with a touch of pink near the base, are held in densely bristled clusters.

Roundheaded Lespedeza is a nutritious plant for grazing livestock; its seeds provide food for birds and other wildlife.

habitat . . .
PRAIRIE. upland mesic-dry, mesic-dry, open woodlands, thickets and disturbed soils
•zone—4a

form . . .
UPRIGHT. slender, single, bushy stem with fine silvery hairs (2–4′)

foliage . . .
ALTERNATE. simple, TRIFOLIATE, leaflets (3), narrow-oblong, grayish green leaves (1–1½″) with entire margin; stalkless, hairy
•color (fall) tan-grayish MAROON
•season—perennial, herb (July–Sept.)

flower . . .
CLUSTER. glabrous, dense, yellow (½″), crowded AXILLARY flower head with purplish tinge; terminal (racemes)

fruit . . .
POD. tiny, tan-brown (½–1″), one-seeded pod
•season—maturing late autumn

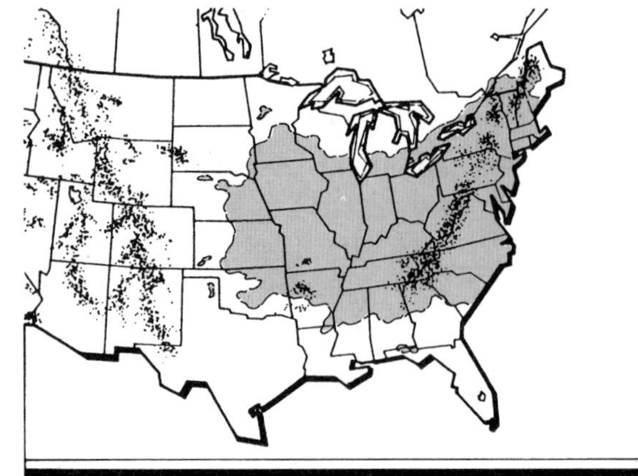

Lespedeza capitata · Roundhead Lespedeza

Rough Gayfeather, or Rough Blazingstar, is common to dry prairies and open woods. It is an erect perennial that rises from a corm or bulb. This ruggedly handsome flower blooms with button-like, lavender-pink florets from August to October. When dried just prior to opening, the flowers are effective in flower arrangements (Owensby, 1980).

habitat . . .
PRAIRIE. upland mesic-dry, mesic-dry, open woodlands, sandy and rocky soils
- zone – 3b

form . . .
ERECT. stiff, slightly ZIG-ZAG, rough stems with grayish hairs (1–4′)

foliage . . .
ALTERNATE. simple, narrow, lanceolate-linear, ASCENDING bright green leaves (2½–3″) with entire margin; rough, short stalked; upper leaves smaller
- color (fall) gray-BROWN
- season – perennial, herb (Aug.–Oct.)

flower . . .
DISK. globular, purplish to rose-pink (⅜–¾″), SPIKE-like cluster head with short, ARCHING filaments; short-stalked, broadly rounded green bracts with pinkish margin; flower head (18–32) ASCENDING upward

fruit . . .
ACHENES. small, seed-like
- season – maturing late autumn

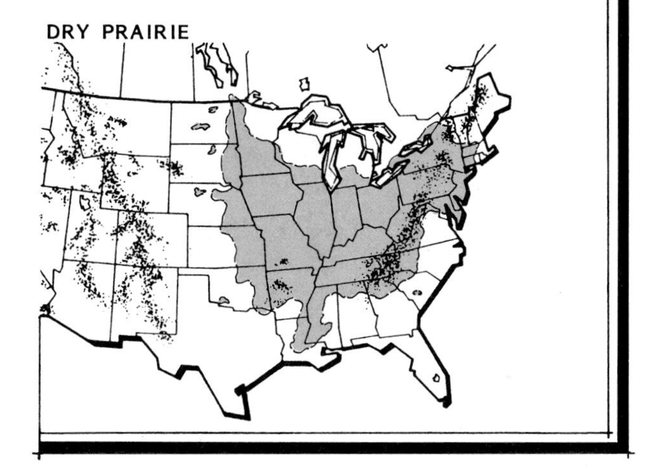

Liatris aspera · Rough Gayfeather

Wildbergamot Beebalm is a perennial forb that grows in dense colonies on rich soils in open woods, roadsides, and old pastures. It is a member of the Mint Family and has that family's characteristics of opposite leaves, square stems, and minty aroma. Wildbergamot's pink flower looks like a ragged pompom; this effect is produced by stamens protruding from many individual flowers that are clustered together.

Indians used Wildbergamot's leaves for tea and to treat fevers, sore throats, colds, headaches, and skin eruptions (Runkel and Bull, 1979)

habitat . . .
PRAIRIE, SAVANNA. upland dry, open woodlands, thickets, open fields and disturbed sites
•zone – 3a

form . . .
UPRIGHT. slender, open branched, 4-angled, reddish brown stem (2–4′)

foliage . . .
OPPOSITE. simple, narrow, lanceolate dull grayish green leaves (2–3½″) with coarse-toothed margin; long petioled with finely matted hairs (velvety)
•color (fall) gray-BROWN
•season – perennial, AROMATIC herb (June–Sept.)

flower . . .
CLUSTER. large, pink to purple (1–1½″), dense, tubular, 2-lipped, showy clusters; terminal with flower bracts

fruit . . .
POD. small, tan-brown (⅛″) seeds
•season – maturing late autumn

DRY PRAIRIE

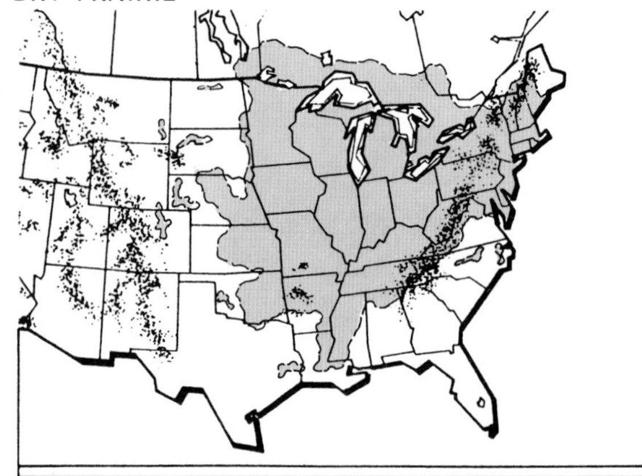

Monarda fistulosa · Wildbergamot Beebalm

Dyersweed Goldenrod is a perennial that grows in dry, upland prairies and open woods. It is often found in sandy and rocky soil. It blooms from August to October. The flower is a one-sided yellow plume that gives the plant an arched appearance. The leaves are also unusual, having small leaflets at the base of larger leaves.

Dyersweed, one of the smallest of the goldenrods, seldom reaches 24 inches in height. Yet it is one of the plants most tolerant of the dry conditions that are typical of upland prairies.

habitat . . .
PRAIRIE. upland mesic-dry, upland dry, open woods and sandy rocky soils
•zone – 3a

form . . .
UPRIGHT. slender clumps of slender, gray downy stems (1–3′)

foliage . . .
ALTERNATE. simple, narrow, oblanceolate leaves (1–2½″), entire margin with small leaflets in leaf axils; dense-haired; lower leaves long-stalked, bluntly toothed
•color – green
•season – perennial, herb (Aug.–Oct.)

flower . . .
INFLORESCENCE. singular, yellow (1½–5″) SPIKE-like plume with spikelets on upper side of curved-ARCHING stem; terminal

fruit . . .
ACHENE. tiny, brown (1/16″) seed with short, fuzzy hairs
•season – maturing autumn

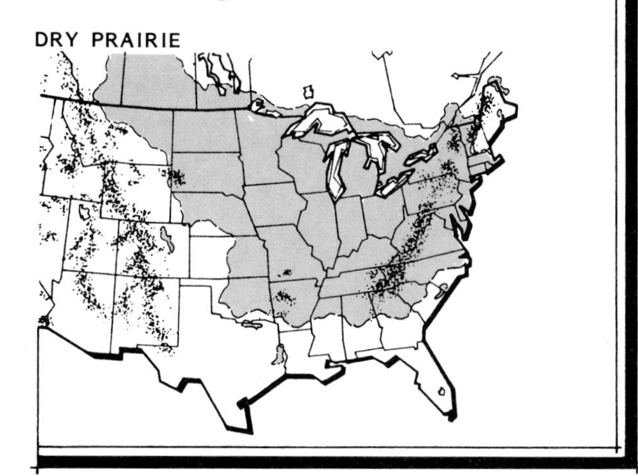

Solidago nemoralis · Dyersweed Goldenrod

Woolly Verbena, or Hoary Vervain, is a perennial flower. It grows in a variety of soils and is one of the most common invaders of overgrazed pastures, but it does not compete well in vigorous stands of native grasses. The name, Woolly Verbena, comes in part from the very dense, velvety covering of whitish hairs on the stem and along the underside of the leaves.

The flowers are lavender or rosy pink and appear in a ring halfway down the flower stalk. The individual flowers, although small, are nearly twice the size of Blue Verbena's flowers.

habitat . . .
PRAIRIE. upland dry, open woodlands, roadsides and disturbed soils
•zone—3a

form . . .
ERECT. slender, sparingly branched, SQUARISH stems with whitish hairs (1–4′)

foliage . . .
OPPOSITE. simple to whorled, ovate-oblong, bright green leaves (2–3″) with coarse-toothed margin; stalkless, thickly clothed with whitish, VELVETY hairs
•color (fall) gray-BROWN
•season—perennial, herb (July–Sept.)

flower . . .
SPIKE. small, bluish pink to purple (1/4″), narrow, tubular clusters, PENCIL-like(8″) elongating spikes; flower head ASCENDING upward

fruit . . .
NUT. flat, tan-brown (1/8″), oblong seeds
•season—maturing autumn

DRY PRAIRIE

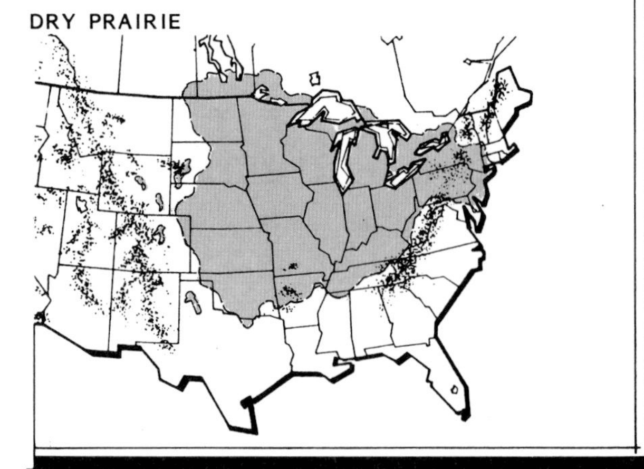

Verbena stricta · Woolly Verbena

The Marsh-Pothole prairie community is found on poorly drained soils of lowland and backwater bogs, or pothole depressions, of the mesic prairie. These areas are frequented by a select group of forbs and grasses. Plants in this elite group must survive a very demanding initiation of several weeks of standing water.

Standing water has a great effect on most plants: they just stop growing. The closer one gets to an area of standing water in the prairies, the fewer species one finds. Pothole depressions are generally composed of a course outer ring of Prairie Cordgrass mixed with other species, such as Swamp Milkweed and Western Ironweed. The centers of these potholes have one main species, Pond Smartweed, which is occasionally intermixed with Common Arrowhead and Common Cattail.

The marsh-pothole prairie has always been a community in transition. The large amount of organic matter produced by tall grass-like plants, such as Prairie Cordgrass, sedges, and bullrushes, eventually fill in marshes and potholes. The whole water system then simply moves over to the next low spot and repeats itself.

This process helped produce the humus-rich soil of the Midwest that now sustains corn and soybeans. Much of the cropland of the Midwest used to be pockmarked with pothole prairies; it is now tillable land because of the active use of farm tile to drain standing water.

Marsh-Pothole Prairie Community

Grasses:

Forbs:

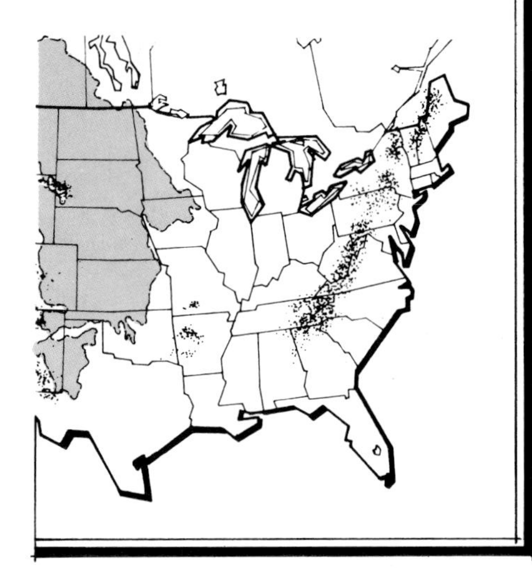

Sedges are common plants in marshes, bogs, wet meadows, pond margins, and wet areas in roadside ditches.

They are very slender and grass-like and are often confused with the Cocklebur because of their cluster of brown seed capsules, clinging high on the stem.

One easy way to identify sedges is by their stems. They are triangular in cross section and solid between joints.

habitat . . .
PRAIRIE. lowland wet-mesic, lowland wet, wet meadows, bogs and pond edges
- zone – 2b

form . . .
UPRIGHT. erect, slender, often 3-ANGLED stems (CULMS) (1½″–3′)

foliage . . .
ALTERNATE. narrow, GRASS-like, green RECURVING leaf blades (12–18″) with sheaths at base clasping stem; 3-RANKED, rolling more or less when dry
- color (fall) gray-BROWN
- season – perennial, herb (June–Sept.)

flower . . .
SPIKE. (female) enclosed in thin SAC (perigynium), bronze-brown (⅛–¼″), globular dense spikes; (male) sessile, symmetrically settled in axil of leaf bracts
- sex – monoecious

fruit . . .
ACHENE. obovoid, tan-brown (⅛–¼″), seed-like, LENS-shaped, nearly black, rough margined
- season – maturing early autumn

MARSH-POTHOLE PRAIRIE

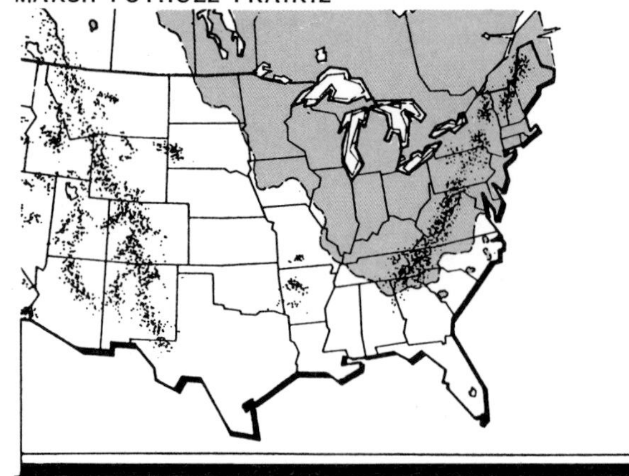

Carex species · Sedges

Scouringrush is found in wet prairies, wet woodlands and along streams and ditches. The erect, unbranched, leafless stems are hollow and jointed. Atop the stems are conical heads that produce spores. The leafless green stems perform photosynthesis for the plant.

"The name, Scouringrush, comes from the use of the plant by Indians and pioneers to scour pots and utensils. The silica embedded within the cell walls of the stem makes Scouringrush very useful for that purpose" (Runkel and Bull, 1979).

habitat . . .
PRAIRIE, SAVANNA. lowland wet-mesic, lowland wet, moist meadows, damp shaded slopes, stream bank and pond margin
- zone—3a

form . . .
ERECT. slender, cylindrical, hollow BAMBOO-like rough stems (⅔–5′)
- rhizome—FLESHY, creeping underground, branched, often bearing small tubers

foliage . . .
SOLITARY. seldom branched, having broad ridges with prominent (20–30) ash gray BANDS outlined by dark edges above and below; sheaths cylindrical with sharp-POINTED teeth, brown-edged soon withering, internodes (3–4″) spaced
- stems—fertile and sterile stems alike containing large amounts of SILICA
- color (fall) darkish GREEN
- season—evergreen; THICKET-like growth

fruit . . .
SPORE. black, short-stemmed (⅛–¼″), CONE-like with sharp-pointed tip; PERSISTING

MARSH-POTHOLE PRAIRIE

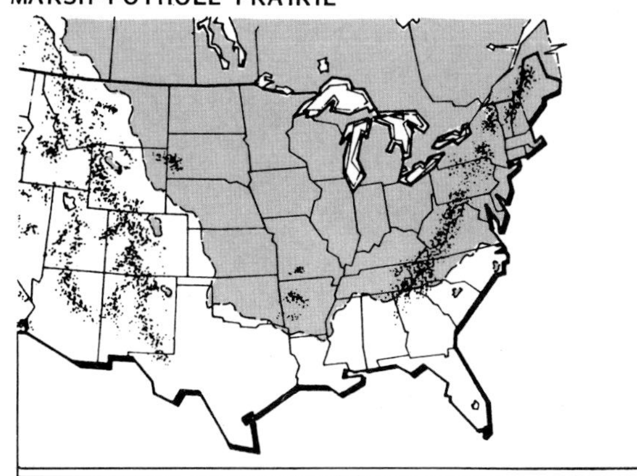

Equisetum hyemale · Scouringrush

Softstem Bulrush is a perennial that grows from rhizomes, or underground stems. It is found in moist swamp meadows, pothole prairies, stream banks, and pond margins. *Scirpus,* or Bulrush, is a cosmopolitan genus of nearly 200 species. Some of the most common varieties grow 5 or 6 feet tall, usually topped with a cluster of many seed capsules.

Bulrush leaves are quite wiry, and the stems of Bulrush differ from those of the Sedge Family in that they are round in cross section rather than triangular.

habitat . . .
PRAIRIE. lowland wet-mesic, wet-mesic, moist swamp meadows, stream banks and shallow pond margin
•zone – 3a

form . . .
ERECT. upright, slender TERETE culms (2–8′)
•rhizome – slender, creeping rootstocks

foliage . . .
ALTERNATE. few basal sheaths, RUSH-like, dark gray-green, (½″) wide stems with brown, sponge-like pith
•color (fall) green-BROWN
•season – perennial (June–Aug.)

flower . . .
INFLORESCENCE. small, greenish brown (⅜″) dense spike; spikelets (5–10), flower heads (umbel); PERFECT

fruit . . .
ACHENE. ovoid, yellowish brown with downward BARBED bristles
•season – maturing late summer

MARSH-POTHOLE PRAIRIE

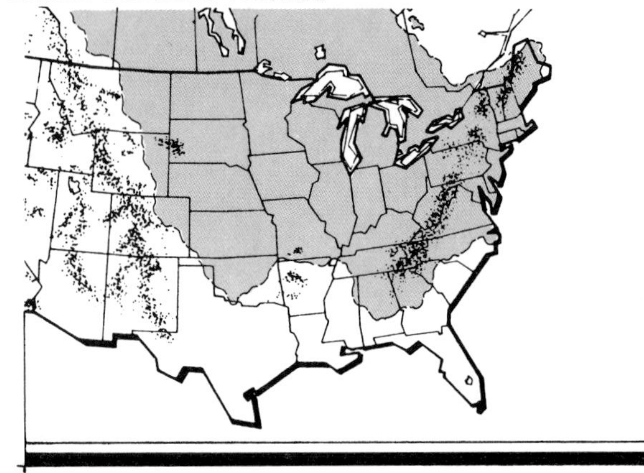

Scirpus validus · Softstem Bulrush

Prairie Cordgrass is a tallgrass that grows in wet places. It is an indicator of land with a high water content and poor soil aeration (Weaver, 1968). Prairie Cordgrass's rank growth and dense, sod-forming habit frequently excludes all other vegetation. The coarse leaves of Prairie Cordgrass with fine teeth along the margins are very much like saw blades, giving reason for one of its common names, Ripgut.

Pioneers and Indians harvested the grass for early hay and later used the mature grass for thatching roofs on permanent lodges (Phillips Petroleum Company, 1963).

habitat . . .
PRAIRIE. lowland wet-mesic, lowland wet, stream and pond margin
•zone—2b

form . . .
UPRIGHT. erect, rigid REED-like culms (3–6′)
•rhizome—rough, scaly

foliage . . .
ALTERNATE. narrow, rough, GRASS-like leaf blade (3–4′) with edges prominently veined; SCABROUS margins often becoming rolled INWARD with long slender point
•color (fall) gray-BROWN
•season—perennial (July–Sept.)

flower . . .
INFLORESCENCE. tan-brown (2–4″) ONE-sided elongated FLATTENED spike in rows of 2s; spikelets (5–20) spirally arranged on stems; ASCENDING

fruit . . .
ACHENE. small, seed-like, often 3-sided
•season—maturing early autumn

MARSH-POTHOLE PRAIRIE

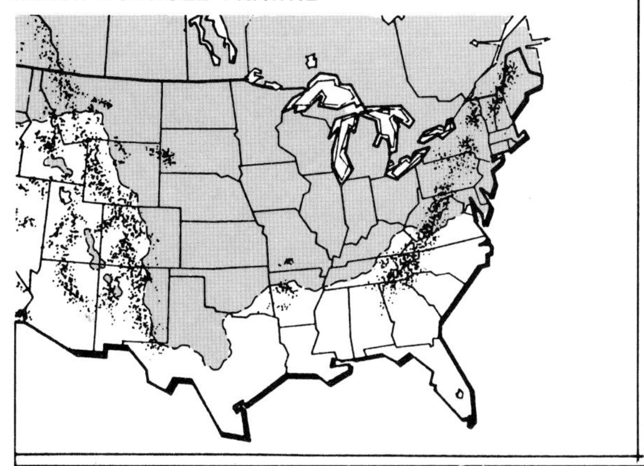

Spartina pectinata · Prairie Cordgrass

Common Cattail grows in rich, wet soils of ponds and lake margins. It blooms from May to July. Each vertical flower stalk is divided into two sections. The upper section contains the male flowers, and the lower, the female flowers. After pollen is shed, the upper section dries up and disappears.

Almost every part of Common Cattail is edible, from the roots to the seedheads. The rootstock was ground into meal by Indians. Young shoots can be eaten like asparagus, and the immature flower spikes can be boiled and eaten like corn on the cob (Runkel and Bull, 1979).

habitat . . .
PRAIRIE. lowland wet-mesic, lowland wet, moist meadows, stream banks and pond margins
•zone – 3a

form . . .
UPRIGHT. slender, stout stemmed, often in dense, erect clumps (4–8′)
•rhizome – creeping (EDIBLE) rootstock

foliage . . .
ALTERNATE. linear, broad (¾–1″) wide leaf blade with SHEATHED base; often exceeding flower head
•color (fall) tan-BROWN

flower . . .
INFLORESCENCE. dense, tan-brown (3–12″), cylindrical CONTIGUOUS spikes, becoming a DOWNY mass
•sex – monoecious

fruit . . .
ACHENE. small, WHITISH cream colored with minute seeds; hairs tightly compressed, short
•season – maturing late summer, PERSISTING through autumn

*Typha angustifolia (Narrowleaf) having narrower leaf blades with SEPARATED male-female flower head

MARSH-POTHOLE PRAIRIE

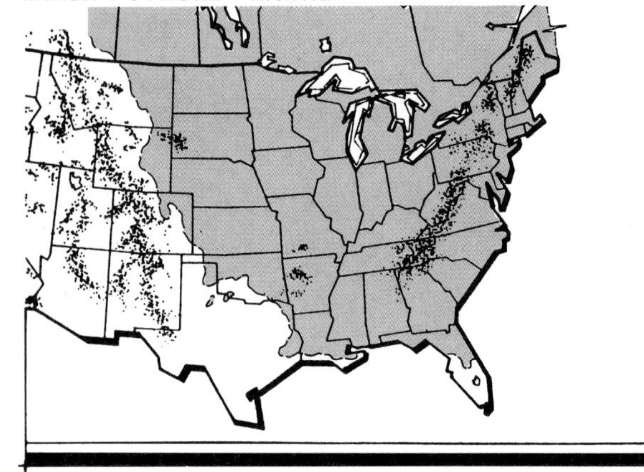

Swamp Milkweed is a handsome perennial that grows in moist meadows and marshes and along lake shores. Its attractive rose-purple flowers bloom in showy dense clusters from May to June.

Milkweeds are usually characterized as having opposite leaves and a milky juice. Swamp Milkweed has opposite lance-shaped leaves, but the stem does not exude as much milky juice when broken as do other milkweeds that grow in drier locations.

habitat . . .
PRAIRIE. lowland wet-mesic, lowland wet, moist meadows, stream banks and disturbed pond margin
- zone – 3a

form . . .
UPRIGHT. erect, open stem with thick, MILKY juice (2–4″)

foliage . . .
OPPOSITE. simple, narrow lanceolate-ovate bright green leaves (3–5″) with entire margin; pubescent beneath, short petioled
- color (fall) gray-BROWN
- season – perennial-herb (May–June)

flower . . .
CLUSTER. small, rose purple to FLESH-color (1/4″) UMBELS in rounded, dense clusters; terminal, FRAGRANT

fruit . . .
POD. elongated, tan-brown (2–4″) seed pods; opening along one side, releasing seeds with silky, WHITE hairs
- season – maturing late summer, PERSISTING through winter

MARSH-POTHOLE PRAIRIE

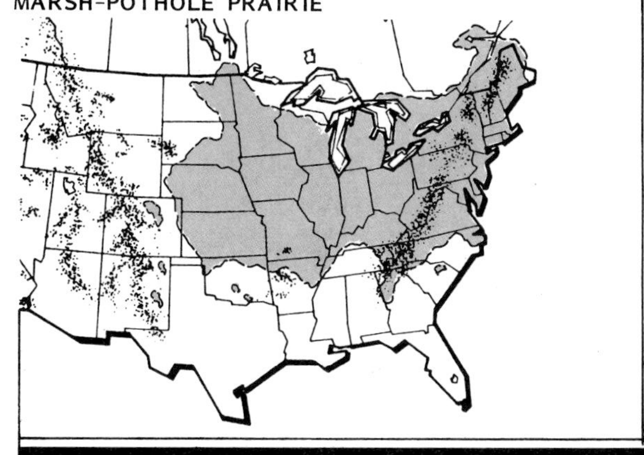

Asclepias incarnata · Swamp Milkweed

Devils Beggarticks are commonly found in the wet soils of stream banks, wide waterways, and marsh edges. Many of the beggartick species have yellow sunflower-like blooms, while others have no ray-flowers at all. Beggarticks are better remembered for the nuisance of their clinging seeds than for the beauty of their petals. The seed is flat and tipped with two rigid spines. These spines, or barbs, have minute hooks pointing backward that adhere tenaciously to clothing.

Crushed parts of the plant emit a disagreeable odor. It is no wonder that many people regard the *Bidens* species as bothersome weeds.

habitat . . .
PRAIRIE. lowland wet-mesic, lowland wet, meadows, open fields and stream and pond margins
•zone – 3a

form . . .
UPRIGHT. open, many branched, slender PURPLISH stem (1–4′)

foliage . . .
OPPOSITE. pinnately COMPOUND, leaflets (3–5), narrow, elliptic-lanceolate dull green leaves (2–6) with toothed margin; stalkless
•color (fall) bronze-BROWN
•season – annual-herb (Aug.–Oct.)

flower . . .
DISK. small, yellow-orange (½″) flower heads with (5–8) LEAF-like bracts; rays (none)

fruit . . .
ACHENE. flat, tan-BROWN (⅛–¼″) ovoid seed, armed with (½″) barbed AWNS
•season – maturing late summer

MARSH-POTHOLE PRAIRIE

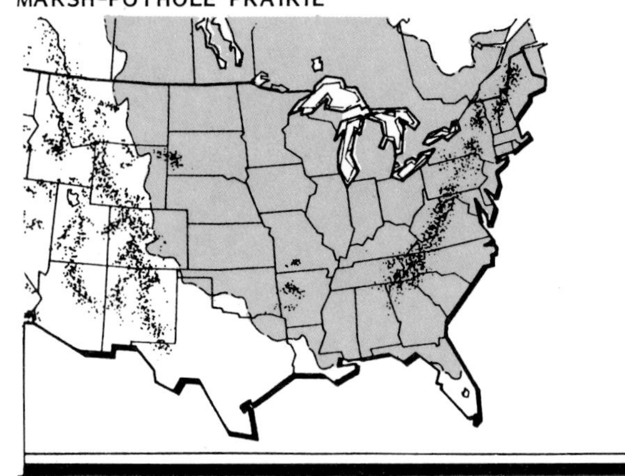

Bidens frondosa · Devils Beggarticks

Bigblue Lobelia is found throughout most of the United States east of the Rocky Mountains. It is common in open woodlands, along streams, and in low, moist areas. The flowers, crowded together on the upper stem, bloom in late summer in the lush growth of shady or half-shaded places. Each flower is split into two lips; the upper lip has two segments, and the lower lip has three.

Some Indians believed this plant to be a secret cure for syphilis; unfortunately, they were mistaken.

habitat . . .
PRAIRIE. lowland wet-mesic, wet-mesic, moist meadows, open fields and disturbed soils
- zone – 3a

form . . .
UPRIGHT. erect to spreading, stout, 4-ANGLED, rigid stems (2–3′)

foliage . . .
ALTERNATE. simple, oblong-lanceolate pale yellow-green leaves (2–6″) with fine toothed-WAVY margin; base sessile
- color (fall) GRAY-brown
- season – perennial-herb (July–Oct.)

flower . . .
SPIKE. erect, purplish BLUE (¾–1″) TRUMPET-shaped, leafy axillary spikes; 2-lipped corolla with lower lip 3-LOBED

fruit . . .
CAPSULE. small, 2-valved, seedy capsule
- season – maturing late summer

MARSH-POTHOLE PRAIRIE

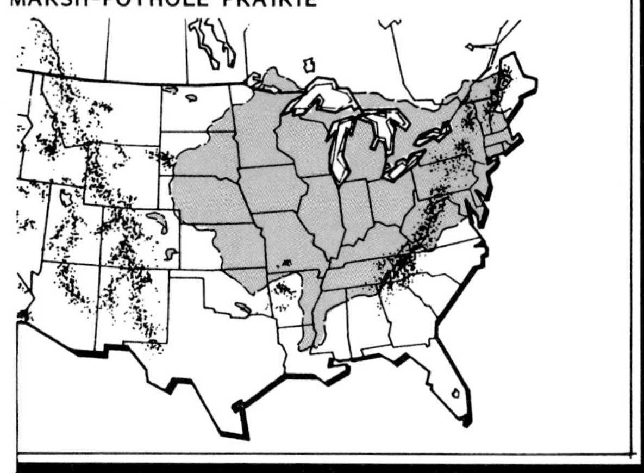

Lobelia siphilitica · Bigblue Lobelia

Pond Smartweed is a common perennial that grows in marshes, along lake shores, and sometimes in shallow water. The smartweed family is easy to recognize; at every point where a leaf is attached, there is a tubular sheath surrounding the stem forming a knot. The family is also called the knotweed family.

The flowers of Pond Smartweed are handsome spikes of scarlet to deep pink.

habitat . . .
PRAIRIE. lowland wet-mesic, lowland wet, moist meadows and disturbed pond margin
•zone – 3a

form . . .
UPRIGHT. open, reddish brown, JOINTED stem with arching branchlets (1–3″)
•rhizomes – thick, black CREEPING rootstock

foliage . . .
ALTERNATE. simple, lanceolate-ovate dark green leaves (3–8″) with entire margin; sheath CLASPING stem at leaf axil
•color (fall) gray-BROWN
•season – perennial (Aug.–Oct.)

flower . . .
SPIKE. slender, rose-PINK to white (3–5″), solitary or in pairs with HAIR-like glands

fruit . . .
ACHENE. small, tan-brown (1/8–1/4″) LENS-shaped; 3-sided
•season – maturing late autumn

MARSH-POTHOLE PRAIRIE

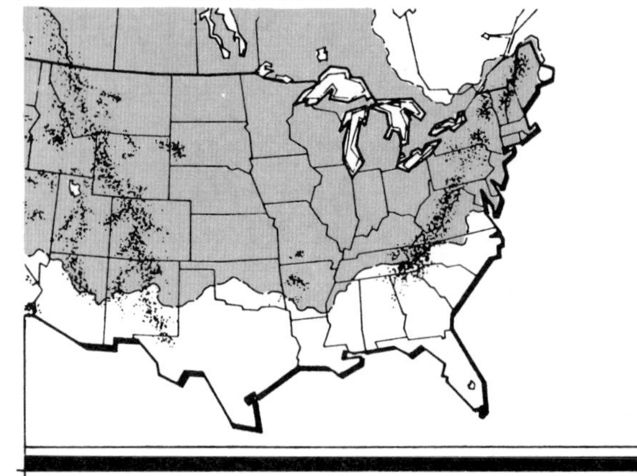

Polygonum coccineum · Pond Smartweed

Common Arrowhead is found in wet bottomlands, marshes, and in shallow, slow-moving water. The name, Arrowhead, comes from the shape of the dark green leaves that are above water. Common Arrowhead also has long, narrow underwater leaves that resemble blades of grass. The flowers appear from July to September. The male flowers are conspicuous and have white petals arranged in whorls of three.

"The plant's underwater tubers were a major food source for many Indian tribes. Explorers Lewis and Clark ate the tubers and referred to them as *wapatoo* in their journals" (Runkel and Bull, 1979).

habitat . . .
PRAIRIE. lowland wet-mesic, wet-mesic, moist meadows, stream banks and swamp-pond margin
•zone–2b

form . . .
ERECT. upright (AQUATIC) slender scape herb (1⅓–3′)
•rhizome–creeping, tuberous-KNOTTED rootstock

foliage . . .
OPPOSITE. simple, ARROW-shaped, long-pointed, shining green leaves (6–12″) with entire margin; long basal lobes
•color (fall) gray-BROWN
•season–perennial (July–Aug.)

flower . . .
DISK. small, yellowish GREEN (½″), whorled (IMPERFECT) (3) white flower heads; stalked-petals

fruit . . .
ACHENE. small (1/16″) dense rounded heads; seeds many (CARPELS) winged
•season–maturing late summer

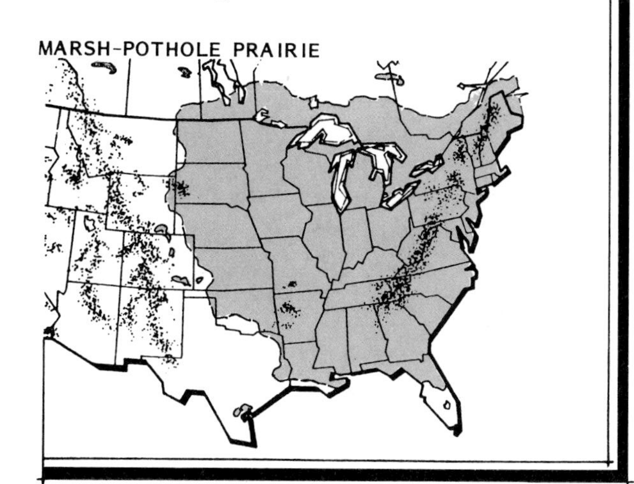

Sagittaria latifolia · Common Arrowhead

Blue Verbena is a common perennial of damp thickets and prairie roadsides. It is a handsome plant with spear-shaped flower spikes. The spikes are numerous and branch upward like arms of a candelabra. Each flower spike has a ring of blue flowers; the flowers at the bottom of the spike bloom first, and the ring of flowers appears to advance upward to the tips of the spike. The leaves are narrow and have prominent toothed margins. In ancient times, the plant was thought to be a cure-all. In fact, its genus name *Verbena* is Latin for sacred plants (Niering and Olmstead, 1979).

habitat . . .
PRAIRIE. lowland wet-mesic, wet-mesic, open moist meadows, roadsides and disturbed soils
•zone–2b

form . . .
UPRIGHT. erect, slender, reddish, 4-SIDED stem (2–5′)

foliage . . .
OPPOSITE. simple, narrow, oblong-lanceolate green leaves (2–6″) with toothed margin; rough lower surface, short stalked
•color (fall) REDDISH green-brown
•season–perennial-herb (June–Sept.)

flower . . .
SPIKE. small, pink-purplish BLUE (1/8″) dense cylindrical (2–4″) PENCIL-like clusters; ascending flower head (5) LOBED

fruit . . .
NUT. small, reddish BROWN (1/16″) nutlets on dense CROWDED spike
•season–maturing early autumn

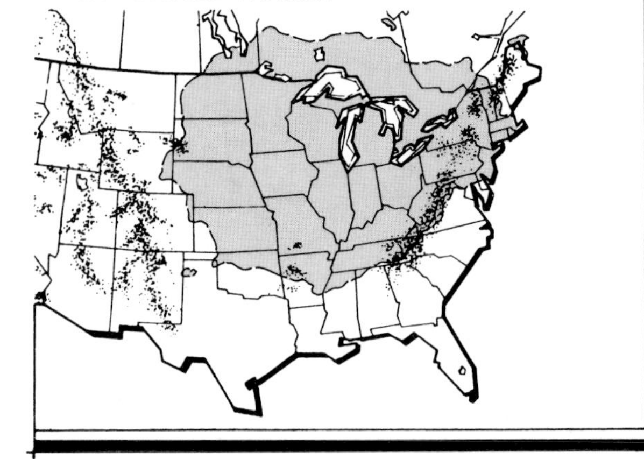

Verbena hastata · Blue Verbena

Western Ironweed is a perennial plant that grows on fertile prairie soils and lowland swales. It blooms in late summer and has attractive floral sprays of clustered purple flowers.

"The name 'ironweed' describes the toughness of the plant's stalk. It is well known as an invader of pastures. It has an extremely bitter taste and is not eaten by livestock" (Phillips Petroleum Company, 1963).

habitat . . .
PRAIRIE. lowland wet-mesic, wet-mesic, moist open meadows, stream edge, roadsides and disturbed soils
•zone – 4b

form . . .
ERECT. upright, stout unbranched stem (2–5′)

foliage . . .
ALTERNATE. simple, oblong-lanceolate dark green leaves (3–6″) with COARSE toothed margin; base sessile, long-pointed, pinnately veined
•color (fall) gray-BROWN
•season – perennial-herb (July–Sept.)

flower . . .
CLUSTER. small, purplish rose-RED (3/8–1/2″), dense FLAT-topped (3–5″) clusters; short stalked flower heads

fruit . . .
ACHENE. small, blunt-pointed (8–10) ribbed
•season – maturing early autumn

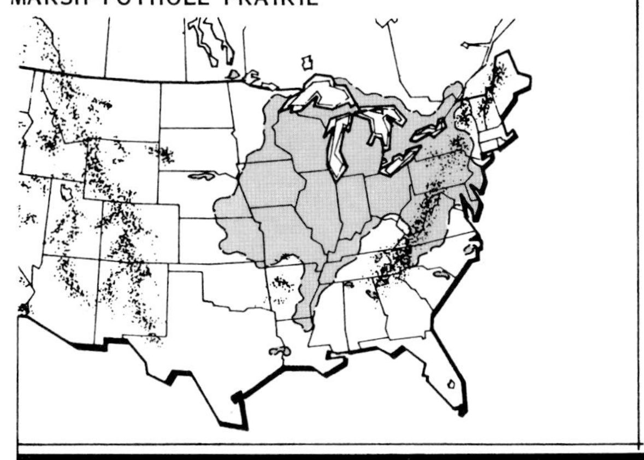

Vernonia fasciculata · Western Ironweed

The Mesic Prairie, or tallgrass prairie, is the community of plants that occupies open spaces that are neither too wet nor too dry. Where optimum conditions exist, many plants are capable of survival; however, only those that are truly competitive last very long in this tight-knit community.

Here, the plan of life is varied. Different plants use soil to different depths, seek light at different heights, and make their maximum demands on the environment during different seasons of the year.

Some plants emerge, flower, and wither away before other species begin their spring growth. Others thrive in the shade and humidity of taller grasses. Many plants shed their lower leaves by midsummer, probably because those leaves are shaded by the rest of the plant. This allows more shade-tolerant species to survive below them.

Plant relationships with their environment are well established on the Mesic Prairie. Compared with single-species agricultural land, the fluctuation of temperature in soil and air is less, the humidity is constantly higher, and evaporation is less; most important, the prairie soil does not wash or blow away.

Grasses:

Shrubs: (less than 6′)

Forbs:

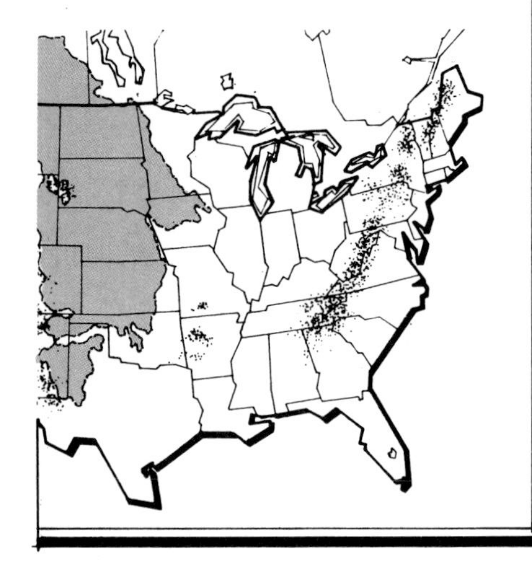

Wet Prairie Community

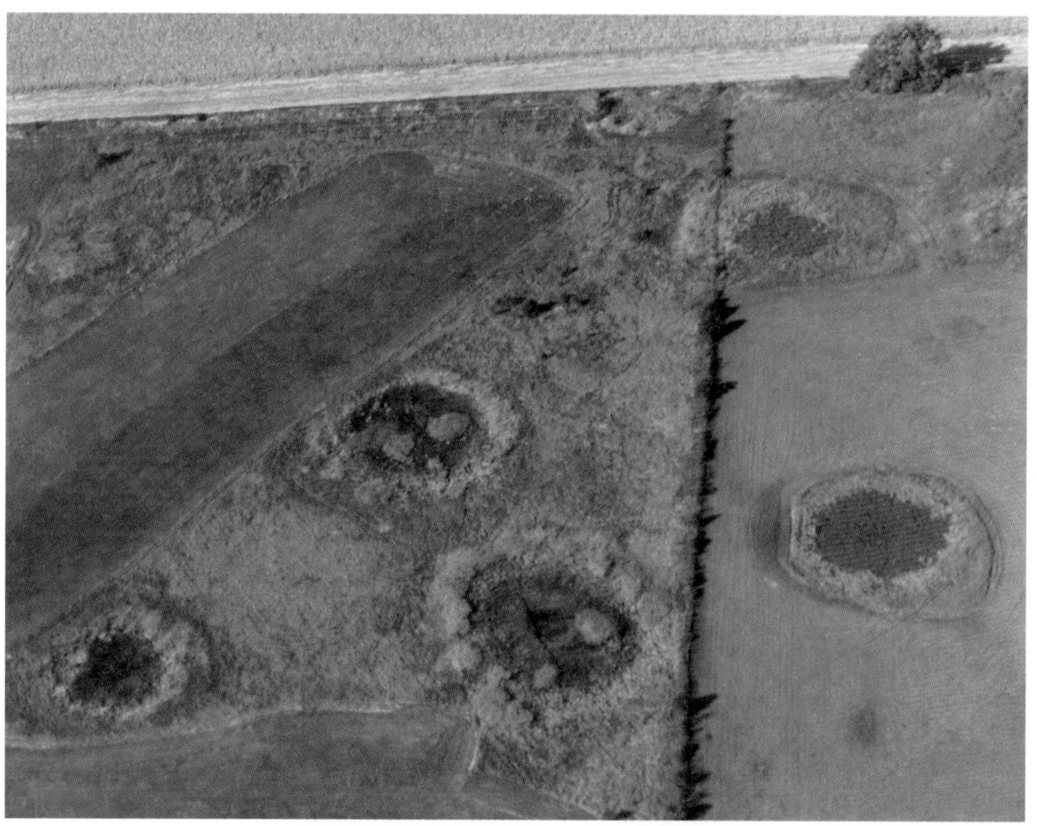

Big Bluestem is a tall perennial grass. Its roots saturate the top 2 feet of soil and may reach depths of 12 feet or more.

Big Bluestem requires more soil moisture than the shorter, more finely rooted Little Bluestem. It generally occupies swales and lowlands, while Little Bluestem dominates the drier knolls.

Big Bluestem grows in close association with two other tallgrasses, Yellow Indiangrass and Switchgrass. Big Bluestem is often called Turkey Foot because its seedhead usually branches into three parts, resembling a turkey's foot.

habitat . . .
PRAIRIE. lowland wet-mesic, mesic-dry, steep rocky slopes and open woodlands
•zone – 2b

form . . .
ERECT. upright, tufted CLUMPS with blue slightly reddish stems (culms) (3–8′)
•rhizome – short rootstock

foliage . . .
BLADE. narrow, green, GRASS-like, lanceolate leaves (8–20″) with FLATTENED base; covered with SILKY hairs
•color (fall) REDDISH maroon-tan
•season – perennial (Aug.–Oct.)

flower . . .
INFLORESCENCE. slender, reddish purple (1–2½″) terminal spikelet clusters dividing into (3) parts; often resembling a TURKEY's foot, turning purple-orange

fruit . . .
GRAIN. small, FUZZY purplish green (⅛–¼″) seed head with SILKY hairs
•season – maturing late autumn

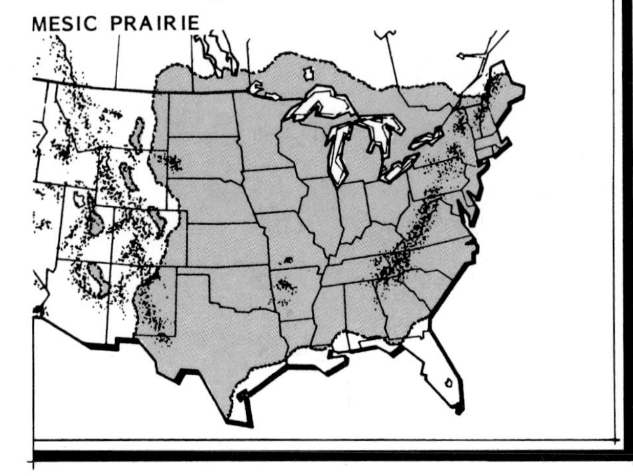

Canada Wildrye is a perennial grass that grows 2 to 4 feet tall. It is winter hardy and begins growth in early fall.

"Canada Wildrye has a high water requirement and produces its best growth in wet areas between stands dominated by Prairie Cordgrass and Big Bluestem, but it also intermingles in both communities" (Weaver, 1968). It occurs in isolated clumps, but is seldom found in pure stands.

Canada Wildrye has an elongated spike-shaped seedhead that causes the stem to bend and droop in a characteristic fashion. Its leaf blades are shaped like spearheads, and has two claw-like ligules that clasp the seed stem.

habitat . . .
PRAIRIE. upland mesic-dry, open woods and stream edge
•zone—2b

form . . .
UPRIGHT. smooth, erect to ARCHING stout stems (2–5′)

foliage . . .
BLADE. broad, flat, green leaf blade (6–12″); SPEAR-like, in pairs; sessile
•color (fall) tan-BROWN
•season—perennial (June–Aug.)

flower . . .
INFLORESCENCE. loose, tan-brown (4–9″), NODDING florets with rigid (3/8″) spikelets; TERMINAL clusters

fruit . . .
GRAIN. tan, oats-like seed with long, bristly AWNS
•season—summer, maturing following summer

(cool season grass)
MESIC PRAIRIE

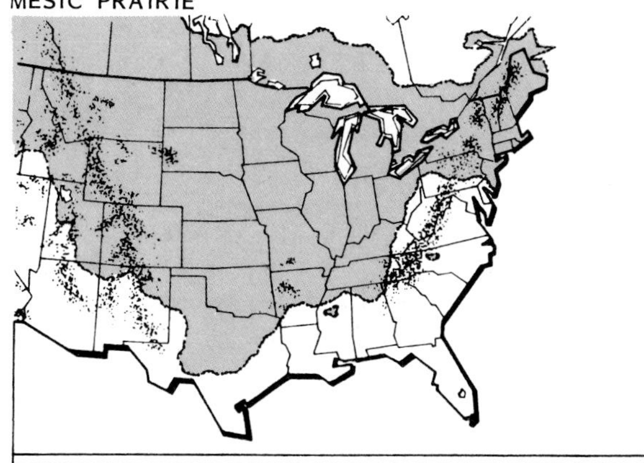

Elymus canadensis · Canada Wildrye

Switchgrass is a tall perennial grass with roots so vigorous that they extend 9 to 11 feet into the soil. "Switchgrass occupies soils that have more moisture than those supporting Big Bluestem and soils that are drier than those supporting Prairie Cordgrass. Switchgrass often forms extensive communities in just such areas" (Weaver, 1968).

Switchgrass has fine-textured, open seedheads. It grows in clumps and can be easily identified by its upright nature and by the prominent nest of hairs where the leaf blade attaches to the stem.

habitat . . .
PRAIRIE. lowland wet-mesic, upland mesic, stream banks and open woodlands
•zone – 2b

form . . .
UPRIGHT. open, stiff, wide-spreading culms (2–6′)
•rhizome – thick, scaly, stout rootstock

foliage . . .
BLADE. narrow, green, GRASS-like, long flat leaves (6–24″) with prominent HAIRS at leaf blade axils; sheathed
•color (fall) orange-PURPLE
•season – perennial (July–Sept.)

flower . . .
INFLORESCENCE. open, reddish PURPLE (1/8–1/4″) floret spikelets in a COMPOUND (6–20″) terminal cluster

fruit . . .
GRAIN. small, tan; SEED-like; glossy
•season – maturing autumn

MESIC PRAIRIE

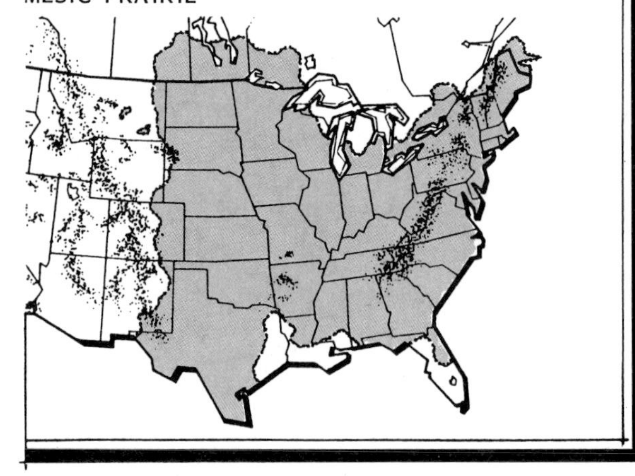

Panicum virgatum · Switchgrass

Yellow Indiangrass is one of the principal grasses of the tallgrass prairie. It is usually found in association with other grasses, and it seldom grows in pure stands. Indiangrass is a perennial that prefers the deep, moist soils of the mesic prairie but will survive on drier sites (Weaver, 1968).

Yellow Indiangrass has a beautiful plume-like seedhead, filled with short, soft, golden brown hairs. Where the leaf blade attaches to the stem, there are two projections, called ligules, which look like rabbit ears. The grass's fall color is deep orange to purple.

habitat . . .
PRAIRIE. upland mesic-dry, upland dry, stream banks, hillsides and open meadows
•zone – 2b

form . . .
UPRIGHT. erect; often forming large-tufted CLUMPS; simple stemmed (culms) (3–8′)
•rhizome – short rootstock

foliage . . .
BLADE. narrow, pale gray-green, GRASS-like, linear leaves (6–20″) with prominent CLAW-like LIGULE at leaf blade axils; sheathed
•color (fall) ORANGE-purple
•season – perennial (July–Sept.)

flower . . .
INFLORESCENCE. narrow, yellow (1/8–1/4″) floret spikelets in a dense (6–12″) shiny PLUME-like mass; fringed with whitish hairs turning bronze-brown

fruit . . .
GRAIN. small; SEED-like with short AWN-like hairs
•season – maturing autumn

MESIC PRAIRIE

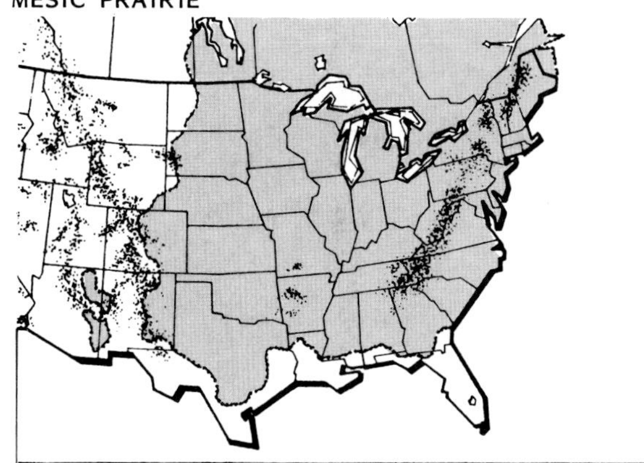

Sorghastrum nutans · Yellow Indiangrass

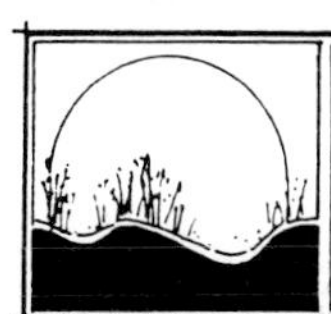

Arkansas Rose is a small shrub that grows in a wide range of soil conditions. It is found in prairies, meadows, and open woodlands. Arkansas Rose blooms in June, and its pink blossoms fill the vicinity with a delicate perfume. After the petals fall, the urn-shaped base of the flower matures into a smooth, red, apple-shaped fruit called a rose hip. "These rose hips are said to contain as much vitamin C as an orange. Indians and pioneers ate the hips, flowers, and leaves, mostly when other food was scarce" (Runkel and Bull, 1979).

habitat . . .
PRAIRIE. upland dry, mesic-dry, open woods edge and roadsides
•zone – 3a

form . . .
ERECT. low to spreading, PRICKLY branched shrub (6–18″)

foliage . . .
ALTERNATE. pinnately COMPOUND, leaflets (7–11), obovate-oblong dark green leaves (1–2½″) with coarse-TOOTHED margin; smooth, glabrous above
•color (fall) MAROON-red
•season – deciduous-perennial (June–July)

flower . . .
CORYMBOSE. large, pink to white (1–2½″) with LOBED outer petals; SOLITARY (2–3) clusters

fruit . . .
HIP. small, red (½″) globular fleshy URN-shaped; many BERRY-like achenes at maturity
•season – maturing early autumn, PERSISTING through winter

MESIC PRAIRIE

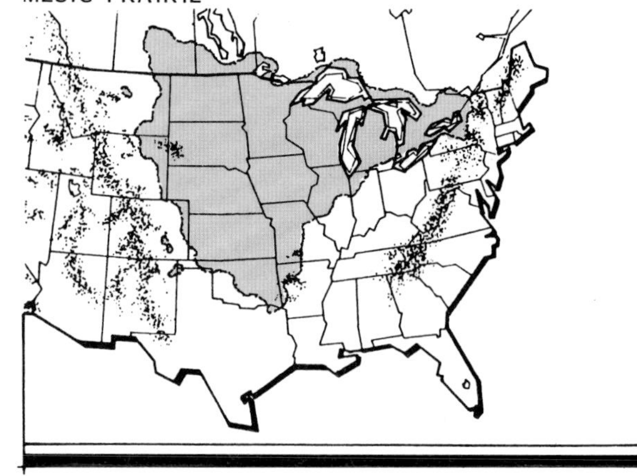

Rosa arkansana · Arkansas Rose

NewEngland Aster is among the largest of the 200 species of asters found in North America; it grows to 6 or more feet. The showy flowerheads are composite, meaning they are made up of many small flowers, called florets. The purple, strap-shaped ray florets surround the tightly packed orange disk florets.

From August to October, fields, swamp edges, and fence-rows are taken over by these robust plants, which produce a profusion of showy flowers. These flowers attract swarms of bees and butterflies. Where they grow amid patches of goldenrod, the color contrast is spectacular.

habitat . . .
PRAIRIE. upland mesic-dry, upland dry, open meadows and stream banks
• zone – 2b

form . . .
UPRIGHT. slender, stout BRISTLY-hair, stemmed (10–36″)

foliage . . .
ALTERNATE. radiating, oblong-lanceolate, dull dark green leaves (2½–4″) with entire margin; upper leaves HEART-shaped with basal lobes CLASPING stem
• color (fall) TAN-brown
• season – perennial-herb (July–Oct.)

flower . . .
DISK. showy, yellow (⅜″) with PURPLISH blue (½–¾″) lanceolate (15–30) rays; flower heads spreading with greenish CURVED bracts; corymbose branched

fruit . . .
ACHENE. small, one-seeded
• season – maturing autumn

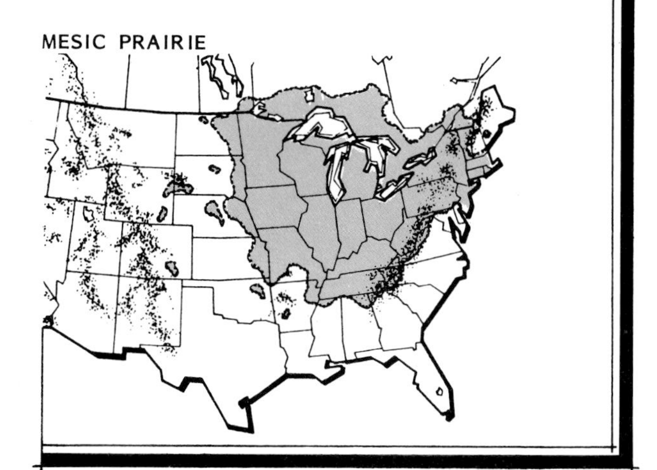

Aster novaeangliae · New England Aster

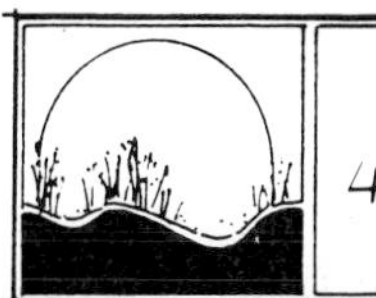

Canada Tickclover is a perennial legume that grows in damp meadows, moist thickets, and ditches. Like most legumes, Tickclover has pea-like flowers and compound leaves in groups of three, and it produces bean-shaped pods. The flowers are rosy purple and bloom from June to September. There are numerous flowers at the tips of the upper branches, but only a few are in bloom at any one time.

Seedpods, flowers, and buds may all appear on a single stem. The seedpods usually have three to five segments that break easily and cling to animal fur or clothing.

habitat . . .
PRAIRIE. upland mesic-dry, lowland dry, open woods and roadsides
•zone – 3a

form . . .
UPRIGHT. slender, erect hairy stems, often bushy appearance (2–6′)

foliage . . .
ALTERNATE. pinnately COMPOUND, leaflets (3–5), oblong-ovate green leaves (2–3″) with entire margins; stipules at leafstalk's base lanceolate
•color (fall) tan-BROWN
•season – perennial-herb (July–Aug.)

flower . . .
CLUSTER. small, PINK-purple (½″) dense PEA-like terminal clusters

fruit . . .
POD. flat, tan-brown, (1″) deeply LOBED, (3–5) jointed (loments) with minute, HOOKED hairs; seeds reddish brown
•season – maturing late summer

MESIC PRAIRIE

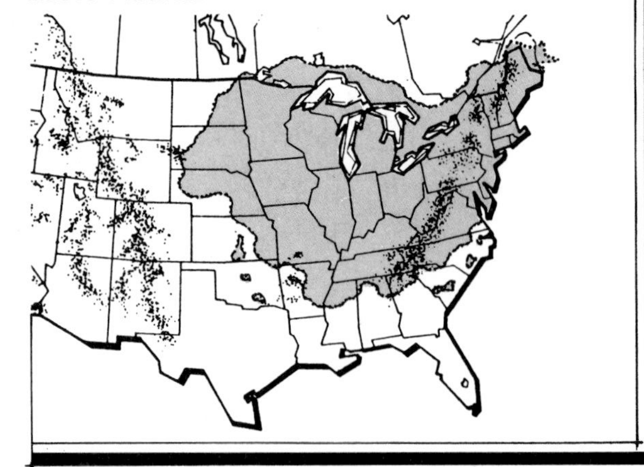

Desmodium canadense · Canada Tickclover

Buttonsnakeroot Eryngo, or Rattlesnake Master, is a perennial with an interesting white flowerhead that resembles a thistle and smells like honey. The yucca-like leaves have spiny edges that make the plant unpalatable to livestock. The plant was once credited with various curative powers and was used as a serum for rattlesnake bite. Rattlesnake Master usually blooms as a companion to Compassplant in July and August.

habitat . . .
PRAIRIE. upland mesic-dry, lowland dry and open woods
•zone – 4b

form . . .
ERECT. stiff to upright, rigidly branched with smooth, BLUISH gray stem (2–6′)

foliage . . .
BLADE. twisting, linear, long, SHARP-pointed blue-gray leaves (6–36″) with bristly, SPINY-edged margin; parallel-veined with leaf sheath CLASPING stem
•color (fall) bluish GRAY
•season – perennial-herb (July–Aug.)

flower . . .
CLUSTER. small, GREENISH white-blue (3/4″), dense florets in ovoid TEASEL-like heads; groupings of 3s
•sex – monoecious

fruit . . .
ACHENE. small, tan-brown, SPINY
•season – maturing late summer

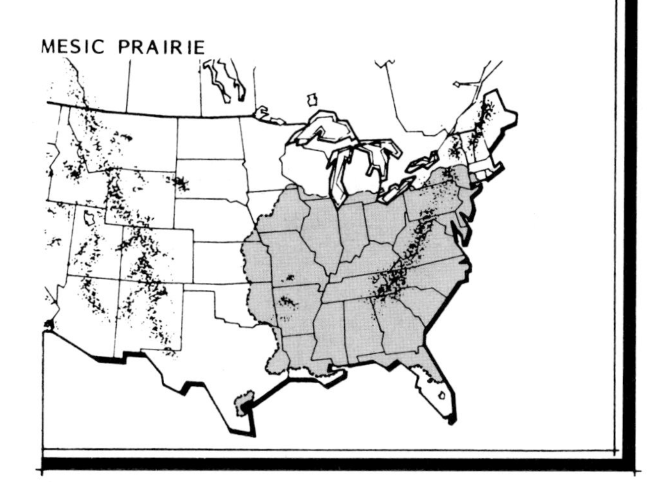

Eryngium yuccaefolium · Buttonsnakeroot Eryngo

Andrews Gentian, also known as Closed or Bottle Gentian, grows in rich damp soils of low-lying thickets and woodland edges. It blooms from late August until there is frost. When in full bloom, the bottle-shaped flower looks like a bud about to open. Only the persistent bumblebee can force its way into such a bloom to gather nectar and pollen.

habitat . . .
PRAIRIE. upland mesic-dry, upland dry, open meadows and roadsides
- zone – 3a

form . . .
ASCENDING. upright to spreading, slender REDDISH stem (1–2′)

foliage . . .
OPPOSITE. narrow, purplish cast, ovate-lanceolate shiny green leaves (2–4″) with entire margin; sessile to whorled below flowerhead
- color (fall) red-MAROON
- season – perennial-herb (Aug.–Oct.) (among the last to bloom)

flower . . .
CLUSTER. cylindrical, PURPLISH blue (1–1½″), dense BOTTLE-like clusters; lobes (5) with whitish band, fringed; nearly closed at top

fruit . . .
CAPSULE. small, dry (1/16″) upright capsule with many seeds
- season – maturing late autumn

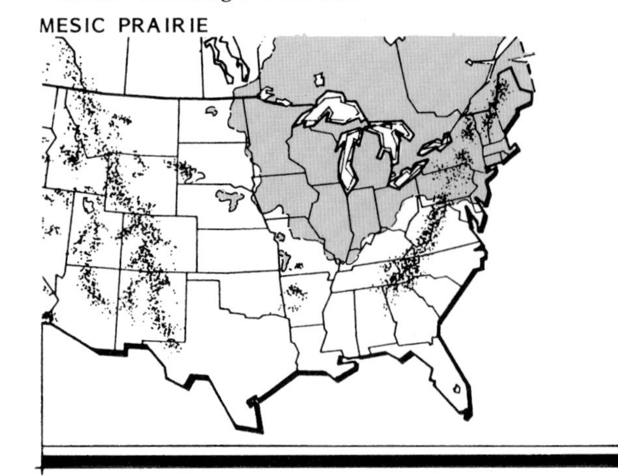

Gentiana andrewsi · *Andrews Gentian*

Sawtooth Sunflower is one of the tallest sunflowers, growing from 5 to 9 feet. It is a perennial that prefers the wet bottomlands and damp pothole regions. Like most plants in its family, Sawtooth Sunflower is best known for its showy yellow flowers. It differs from other sunflowers, however, in that both its disk and ray-flowers are yellow; the disk-flowers of other sunflowers are usually brown.

Sawtooth Sunflower's leaves are very distinctive. They have coarse teeth that gives purpose to the name *Grosse-serratus*, or Sawtooth.

habitat . . .
PRAIRIE. upland mesic-dry, lowland wet-mesic, open meadows and roadsides
•zone – 3b

form . . .
UPRIGHT. erect, smooth stem, often waxy REDDISH brown (6–10′)
•rhizome – elongated rootstock

foliage . . .
ALTERNATE. simple, oblong-lanceolate pale green leaves (3–10″) with coarse SAW-toothed margin; BOAT-shaped with slender petiole; hairy beneath
•color (fall) RED-brown
•season – perennial (Aug.–Oct.)

flower . . .
DISK. large, yellow (1–3″), dense, CONE-like disk florets with yellow (10–20) rays; terminal clusters

fruit . . .
ACHENE. small, tan-brown (1/8″), flat; seed-like
•season – maturing late summer

MESIC PRAIRIE

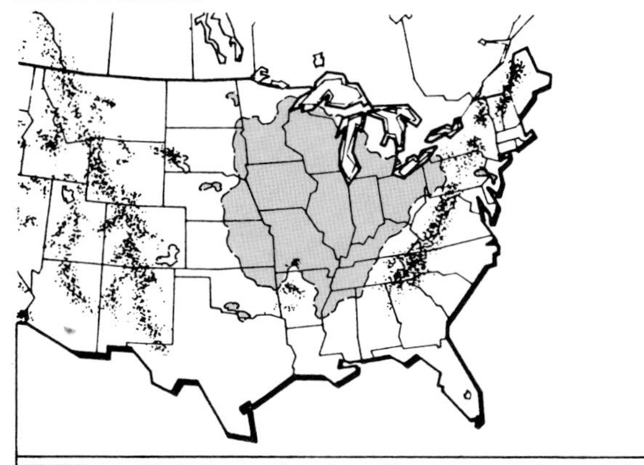

Helianthus grosseserratus · Sawtooth Sunflower

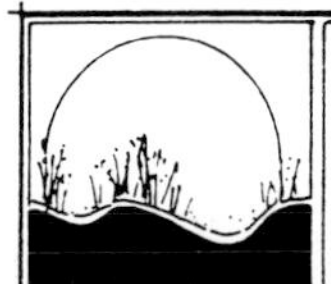

Kansas Gayfeather, or Prairie Blazingstar, is an attractive perennial with a single stalk rising from a woody corm. From July to September, it fills wet prairies and meadows with purple blooms. Its scientific name describes the flower: *Pycnostachya* means "with crowded spike." The many narrow leaves are smaller at the top of the stem and larger at the base. This plant is one of the most popular of the blazingstars and has great appeal, whether blooming in a prairie setting or in a perennial border.

habitat . . .
PRAIRIE. upland mesic-dry, lowland wet-mesic, open meadows and roadsides
•zone–3b

form . . .
ERECT. upright, slender stemmed with HAIRY leafy stalk (3–5′)
•rhizome–tuberous, CORM-like (woody)

foliage . . .
ALTERNATE. narrow, WHORLED, linear shiny green leaves (4–8″) with entire margin; lower lanceolate-shaped, ASCENDING
•color (fall) gray-BROWN
•season–perennial-herb (Aug.–Sept.)

flower . . .
INFLORESCENCE. small, rose PINK to purple (½″) WAND-like spike (5–18″) dense, disk clusters with short ARCHING filaments; short stalked, sharp-pointed bracts

fruit . . .
ACHENE. small, tan-brown; ANGLED; seed-like
•season–maturing autumn

MESIC PRAIRIE

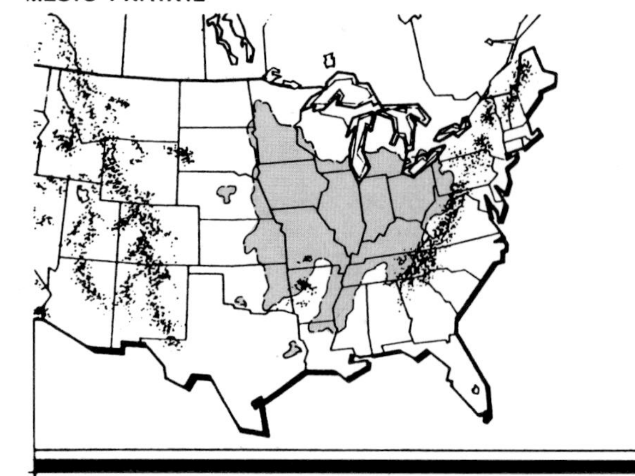

Liatris pycnostachya · Kansas Gayfeather

Purple Prairieclover, a deep-rooted, perennial legume, is most abundant on mesic tallgrass prairie sites. This slender plant has many small T-shaped leaves that have a binding or constipating effect when used for tea. The flower is very attractive, blooming from the base to the tip in a halo fashion, displaying a delightful appearance. Purple Prairieclover, with its thick stands, brings new life to the prairie in June and July with its brightly colored flowers and the activity of the many insects that are drawn to it.

habitat . . .
PRAIRIE. upland mesic-dry, lowland dry, open meadows and roadsides
•zone – 3a

form . . .
ERECT. slender, SILVERY cast stems often forming clumps with branching above (1–3′)
•rhizome – short, vertical rootstock

foliage . . .
ALTERNATE. pinnately COMPOUND, leaflets (3–5), narrow, linear dark green leaves (½–¾″) with entire margin; short stalked, FLATTENED, slightly hairy
•color (fall) GRAY-brown
•season – perennial-herb (June–Sept.)

flower . . .
SPIKE. cylindrical, pinkish PURPLE (½–2″), ELONGATED showy, dense terminal clusters; PERFECT
•sex – monoecious

fruit . . .
POD. erect, silvery gray (½–1½″) with wings or KEELED joint (1–2); seeded
•season – maturing late summer

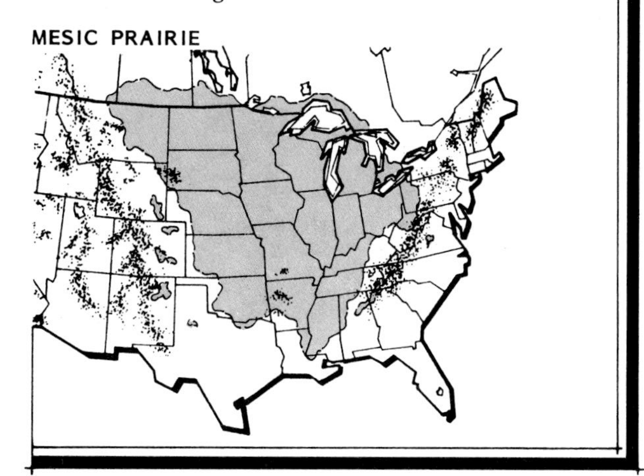

Petalostemon purpureus · Purple Prairieclover

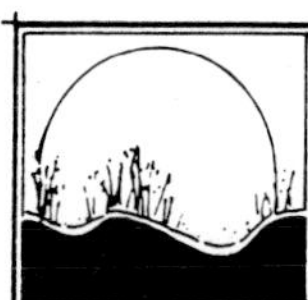

Narrow-leaf Mountainmint is a perennial that, despite its name, grows far from the mountains in open woods, meadows, and prairies. The leaves are numerous, smooth, and very narrow. Narrow-leaf Mountainmint blooms from July to September, and its flowers are white, flat-topped, open clusters.

Typical of plants in the Mint Family, it has opposite leaves, square stems, and a definite spearmint odor when crushed. Its dried leaves make a good tea.

habitat . . .
PRAIRIE. upland mesic-dry, lowland wet-mesic, woods edge and open meadows
•zone – 4a

form . . .
UPRIGHT. erect, densely branched glabrous 4-ANGLED stems (1–3′)

foliage . . .
OPPOSITE. narrow, linear-lanceolate dark green leaves (¾–1½″) with entire margin; SESSILE, glabrous
•color (fall) GRAY-brown
•season – perennial-herb (July–Sept.)

flower . . .
INFLORESCENCE. small, white to LAVENDER (¼–⅜″), dense rounded (2″) clusters, corolla (2) lipped; AROMATIC

fruit . . .
ACHENE. small, gray-brown nutlet; SINGLE-seeded
•season – maturing late summer

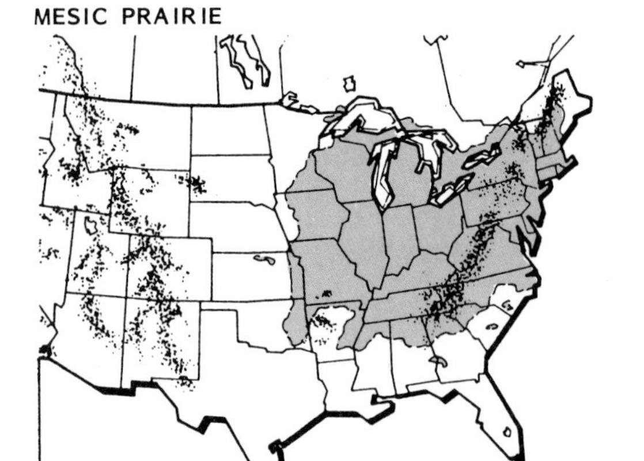

Pychnanthemum tenuifolium · Narrow-leaf Mountainmint

Prairieconeflower, or Grayheaded Coneflower, is a perennial that is tolerant of many conditions. It grows in dry to moist soils in open prairies and at the edges of woods. Its leaves are deeply divided, and they have three to seven coarsely toothed segments.

Prairieconeflower is one of the most obvious prairie plants to bloom along country roads and railroad tracks. Its flowers are very appealing. The drooping yellow rays move with a gentle breeze, covering prairies with golden yellow from June to September. The brown disk florets eventually fall away, exposing the gray conehead that persists throughout the winter.

habitat . . .
PRAIRIE. lowland dry, open meadows and roadsides
•zone – 3b

form . . .
UPRIGHT. erect, slender hairy stem with upper part much branched (3–5′)
•rhizome – woody

foliage . . .
ALTERNATE. pinnately COMPOUND, leaflets (3–7), narrow, oblanceolate green leaves (3–5″) with entire or (2–3) CLEFT margin; FEATHER-like appearance
•color (fall) gray-BROWN
•season – perennial (June–Sept.)

flower . . .
DISK. cylindrical, ash GRAY turning brown (3/4″) knob with yellow (1 1/2″), DROOPING rays; terminal WIRY (5″) stem; PERFECT

fruit . . .
ACHENE. smooth, grayish brown (1/8″), 4-ANGLED on cone; BUTTON-like head (winged)
•season – maturing late summer

MESIC PRAIRIE

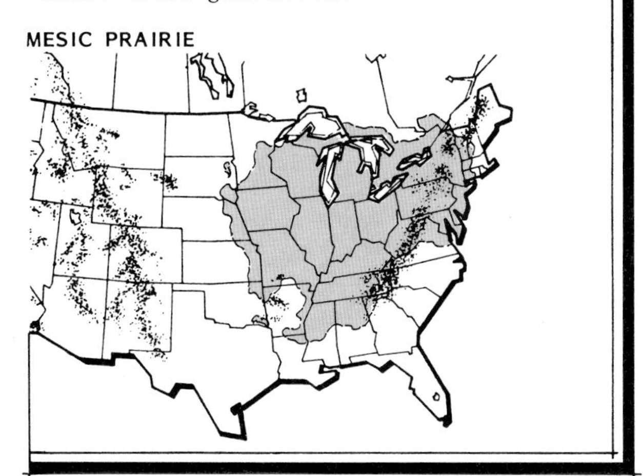

Ratibida pinnata · Prairieconeflower

Compassplant is a famous, deep-rooted perennial of the Midwest tallgrass prairie. It is a plant commonly dismissed as just another sunflower with large yellow ray and disk-flowers. However, it has several unique differences. First, the stem is grizzly-hairy and the large basal leaves are thick with many deep incisions, resembling an oak leaf. In addition, the lower leaves tend to orient themselves in a north/south direction, taking full advantage of the morning and afternoon sun, hence the name Compassplant. Finally, its presence indicates moist, deep subsoil in areas that are not overgrazed or mowed on a regular basis.

habitat . . .
PRAIRIE. upland mesic-dry, lowland dry, open meadows and roadsides
- zone – 4a

form . . .
UPRIGHT. rough, bristly, stiff stemmed, often extruding RESINOUS juice (4–8′)

foliage . . .
ALTERNATE. deep, pinnately LOBED oblong-lanceolate, whitish green leaves (12–18″) with toothed lobes; edges mostly oriented (NORTH-SOUTH) vertical position
- color (fall) gray-BROWN
- season – perennial-herb (July–Sept.)

flower . . .
INFLORESCENCE. several yellow (¾–1″) disk florets with (18–30) yellow rays (1–2″) on short HAIRY stalk
- sex – monoecious

fruit . . .
ACHENE. notched, tan-grayish brown; WINGED or flattened seed
- season – maturing late summer

MESIC PRAIRIE

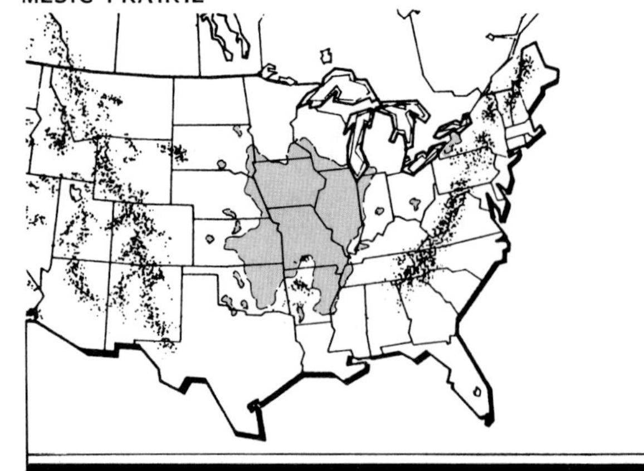

Silphium laciniatum · Compassplant

Stiff Goldenrod is a deep-rooted perennial that is tolerant of a variety of prairie soils. Its name comes from the very stiff, rough-textured leaves that help to retard moisture loss during hot, dry periods. The bloom is a mass of tiny, golden flowers held in a flat-topped cluster atop a rigid, hairy stem.

Goldenrods are an all-American flower; few goldenrod species grow in other countries. This well-known flower group is needlessly feared as the cause of hay fever. The real culprit is ragweed, which releases its pollen at the same time the goldenrods are in bloom.

habitat . . .
PRAIRIE. upland dry, open meadows and roadsides
•zone—3a

form . . .
UPRIGHT. erect, stiffstemmed with DOWNY surface (2–5′)
•rhizome—short, creeping rootstock

foliage . . .
ALTERNATE. simple, oblong-elliptic, rough gray-green leaves (¾–1½″) with rigid margin; ascending with base CLASPING stem
•color (fall) gray-BROWN
•season—perennial-herb (Aug.–Oct.)

flower . . .
INFLORESCENCE. small, yellow (¼″), BELL-shaped, disk florets forming dense, FLAT-topped clusters; large terminal well-DEVELOPED rays
•sex—monoecious

fruit . . .
ACHENE. small, dry (1/16″) with bristly PAPPUS
•season—maturing late autumn

MESIC PRAIRIE

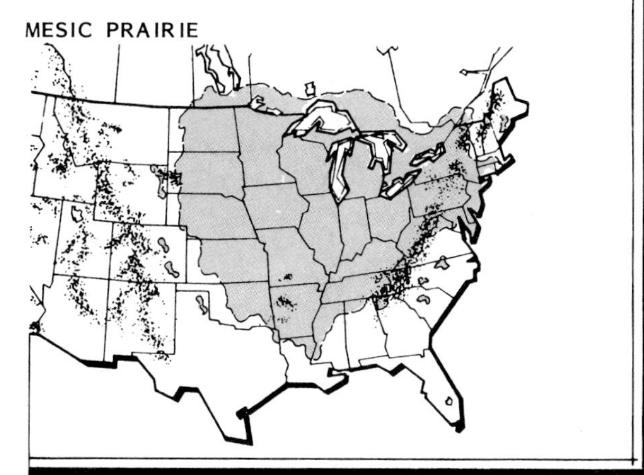

Solidago rigida · Stiff Goldenrod

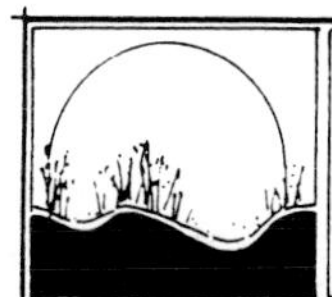

Nodding Ladiestresses is the name of a marsh orchid that grows in damp meadows, moist thickets, and grassy swamps. This little orchid is one of a dozen species of Ladiestresses. It blooms during August in the North and November in the South.

Nodding Ladiestresses has white, fragrant flowers arranged in a spiral on a spike. The scientific name *Spiranthes* means "coil and flower," alluding to the spiral arrangement of the plant's flowers. The fact that the flowers appear to nod makes it different from other species.

habitat . . .
PRAIRIE. upland mesic-dry, lowland wet-mesic, lowland dry and open damp meadows
•zone—3b

form . . .
UPRIGHT. slender, delicate with often TWISTED stem (6–24″)
•rhizome—tuberous

foliage . . .
ALTERNATE. simple (basal) linear-oblanceolate green leaves (6–10″) with upper leaves SCALE-like
•color (fall) gray-BROWN
•season—perennial-herb (Aug. [North]–Nov.[South])

flower . . .
SPIKE. small, creamy white (¼–⅜″), nodding, hooded, PEA-like (3–5″), twisting spike; ascending SPIRAL-like
•sex—monoecious, FRAGRANT

fruit . . .
CAPSULE. small, (3) chambered with many seeds
•season—maturing late summer

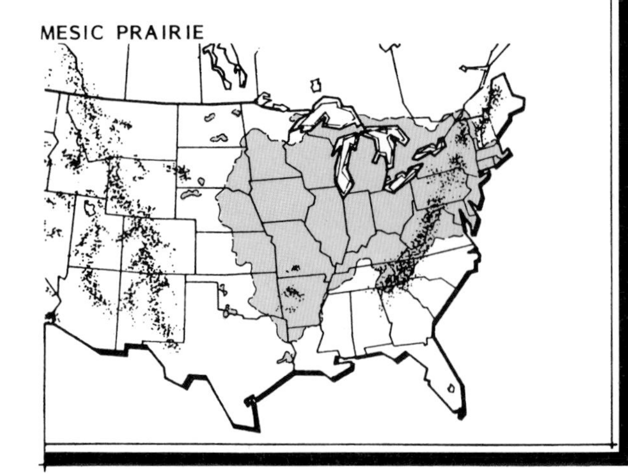

Spiranthes cernua · Nodding Ladiestresses

Culversphysic, or Culversroot, is a tall perennial found along roadsides, in meadows, and in damp, open woods. The flower is a wispy, white bloom that appears from June to September. The leaves surround the stem in whorls of three to nine.

Culversphysic is a flower that resembles veronica and can easily be grown in a wildflower garden. The root of this plant is a useful cathartic and is named after Dr. Culver, who popularized the plant.

habitat . . .
PRAIRIE. upland mesic-dry, lowland wet-mesic, woods edge and open moist meadows
•zone – 3a

form . . .
UPRIGHT. erect, stout PUBESCENT-stemmed (3–6′)

foliage . . .
WHORLED. simple, lanceolate (3–6) green leaves (2–4″) with SHARPLY toothed margin; short-petioled, pubescent beneath
•color (fall) GRAY-brown
•season – perennial-herb (June–Sept.)

flower . . .
SPIKE. small, tubular, white to pale blue (1/4″) dense pyramidal (4–7″) SPIKE-like clusters; terminal

fruit . . .
CAPSULE. small, ovoid-pointed seed
•season – maturing late summer

MESIC PRAIRIE

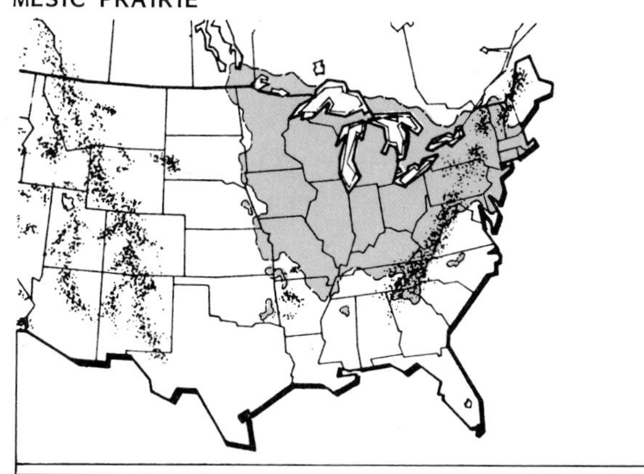

Veronicastrum virginicum · Culversphysic

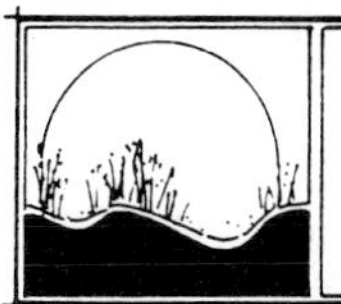

Golden Zizia, or Golden Alexander, is a delicate perennial of moist woods, prairie meadows, and thickets. The light green leaves are compound, usually with three leaflets. The leaflets are sharply toothed, a distinction that separates Golden Zizia from two other members of the Parsley Family, the Yellow Pimpernel and the Meadow Parsnip (Runkel and Bull, 1979).

The blooms of Golden Zizia appear from April to June and attract countless numbers of small butterflies. The golden flowers of this attractive plant are arranged in loose, radiating clusters atop the main stalk.

habitat . . .
PRAIRIE. upland mesic-dry, lowland wet-mesic, open woodlands, moist meadows and thickets
•zone – 3a

form . . .
UPRIGHT. erect, slender glabrous stem (1–3′)

foliage . . .
ALTERNATE. simple, trifoliate COMPOUND, leaflets (3–5), lanceolate, yellowish green leaves (3–5″) with fine-toothed margin; long-POINTED tip
•color (fall) gray-BROWN
•season – perennial-herb (April–June)

flower . . .
UMBEL. small, yellow (1/16″), dense globular clusters in FLAT-topped (2–3″) compound umbels; center umbel stalkless

fruit . . .
CAPSULE. smooth, oblong to flattened (1/8″), NUT-like, (2) seeded capsule; seeds not ribbed
•season – maturing late summer

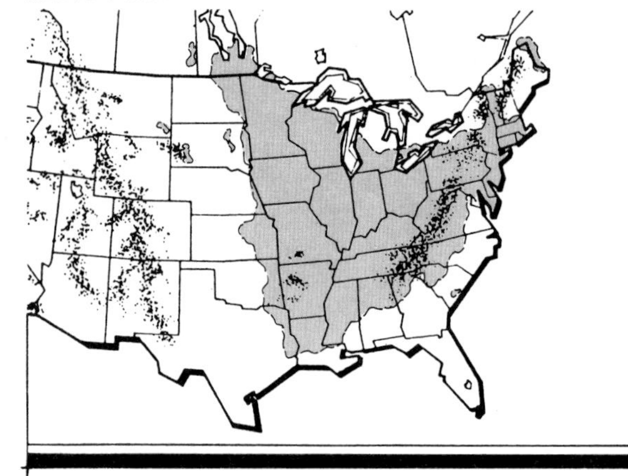

Prior to agricultural development, the midwestern region was predominantly an Oak Savanna community. The savanna, a transition area between grassland and forest, has a parklike quality because of its mixture of grass and trees.

Before agricultural development, this transition area was a battleground between forest and prairie. The savanna could narrow or widen several hundred miles depending on which side was winning at any given time.

The prairie had a great ally in this battle with the forest; that ally was fire. In the spring, before new grasses covered the landscape with unburnable greenery, prairie fires ran rampant. Fire killed most woody plants while leaving the grass unharmed because their sustaining parts, their roots, were protected below ground.

A few woody plants can protect themselves from fire. Sumac and aspen can rejuvenate themselves from underground roots and rhizomes. Bur Oak has an armor plate in the form of a thick, corky bark that covers the tree to the very tips of its branches. According to Aldo Leopold in *A Sand County Almanac* (1977), "Bur oaks were the shock troops sent by the invading forest to storm the prairie; fire is what they had to fight." The widespreading, picturesque crowns of these majestic giants are beautiful reminders of the past battles between prairie and forest before people came to control the situation.

Oak Savanna Community

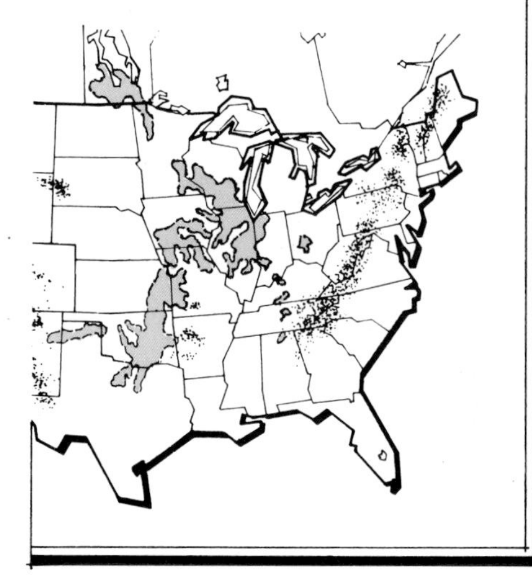

Common Honeylocust has become a common tree in urban settings. It provides filtered shade, and its small leaves are not the nuisance that those of larger-leaved trees can be.

Honeylocust flowers, too, are quite small, in pendant clusters of inconspicuous green. Their fragrance does not come close to the intensity of the Black Locust, but honey bees do not seem to care. They fill the air with the sound of their humming each spring as the honeylocusts come into bloom.

Later the twisted seedpods develop, changing from red-green to a shiny maroon-brown as they reach maturity. When only half-ripe, they are filled with an edible sweet pulp that gives reason for the word honey in this tree's name. The pods are very ornamental and persist after the leaves turn their yellow autumn color and fall to the ground.

The most spectacular features of this tree are its clusters of thorns or spines that jut out from the dark gray bark of the trunk and branches. No other tree is so well armed against invaders. For those who find the thorns too threatening, there are now many thornless varieties available.

Honeylocust, like Bur Oak, is quite tolerant of different soil types and moisture conditions. It grows both at the edges of prairies in dry savanna soils and on floodplains where moisture availability and soil types are quite variable. Honeylocust is a good ornamental that grows fast and is long-lived. It does, however, suffer from mites and Mimosa Webworm invaders.

habitat . . .
FOREST, SAVANNA. upland mesic-dry, lowland wet-mesic, floodplains, open or rocky hillsides and woods edge
- zone – 4b

form . . .
IRREGULAR. globular, open, small canopy tree (50–75′)
- branching – PICTURESQUE, horizontal with coarse, spreading limbs
- twig – shiny, ZIG-ZAG, stout, red-brown, ARMED with 3-branched (1–3″) thorns having small buds hidden by leaf scar; brown (inconspicuous)
- bark – grayish brown-black, shallow-furrowed with long, irregular, vertical PLATE-like ridges, curling along margin; trunk often armed with clustered spines reaching (8′)

foliage . . .
ALTERNATE. pinnately or bipinnately COMPOUND, leaflets (26–32) elliptic-oval, YELLOW-green leaves (¾–1½″) with entire margin; fine-textured, hairless beneath
- color (fall) pale YELLOW
- season – deciduous

flower . . .
SPIKE. small, yellow-green (⅜″), BELL-shaped, pyramidal (4″), spike-like clusters; PENDULOUS
- sex – dioecious

fruit . . .
LEGUME. red-BROWN to purple, twisting (6–18″) BEAN-like pod with sweet (EDIBLE) seeds; pendent
- season – maturing late summer

OAK-SAVANNA COMMUNITY

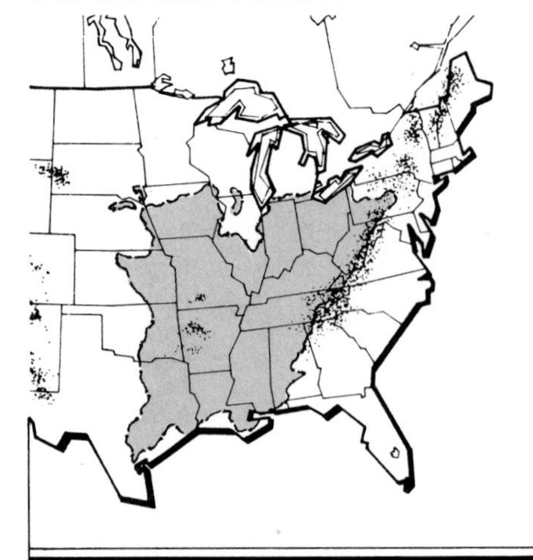

Gleditsia triacanthos Common Honeylocust

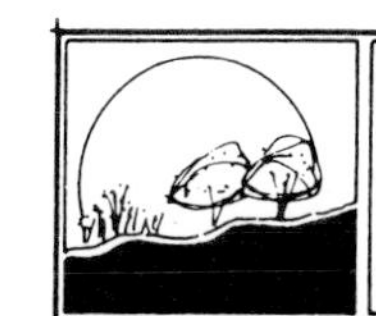

Eastern Redcedar has a very wide distribution range for an evergreen tree, due in part to the dispersion of its fruit by the many birds that prefer it for food. Often a line of Redcedars are soon flourishing along a fencerow, having sprouted from the undigested seeds in bird droppings.

Eastern Redcedar is a common invader of pastures and prairies. Young trees, and occasionally older ones, have two types of leaves. The juvenile form is sharply pointed and needle-like, while the mature form is round and overlapping.

The wood of Redcedar is soft, close-grained, and fragrant. The brownish-red to white, patterned wood is used for paneling, chests, closets, pencils, and many other household items. Despite its softness, Redcedar is remarkably resistant to rot, and it is therefore known as a good material for long-lasting fence posts, shingles, and decks.

Eastern Redcedar is hardy, slow growing, and may live for 300 years. It is a useful tree for screening purposes, in windbreaks, or as a background for showy plants, but it is an alternate host for Cedar-Apple Rust. The rust is an unsightly, red gelatinous mass in spring that forms a hard cone-like cluster in winter, but it does not seem to harm the tree. However, many apple, hawthorn, and mountainash trees suffer rust spots on leaf and fruit when planted in an area with infected Redcedar.

habitat . . .
SAVANNA. upland dry, steep rocky bluffs, dry barren lands and open alluvial woods
- zone – 3a

form . . .
CONICAL. broad, columnar, small canopy tree (50–75′)
- branching – HORIZONTAL to ascending limbs with branchlets drooping
- twig – slender, GREEN to yellow-brown (winter) with 4-ANGLED, minute buds; inconspicuous
- bark – thin, FIBROUS, red-brown, exfoliating; peeling vertically

foliage . . .
OPPOSITE. acicular, SCALE-like, dark GREEN leaves (1/16–3/8″) with fine texture; prickly, NEEDLE-like
- color (fall) OLIVE-green, maroon
- season – evergreen, AROMATIC (young)

flower . . .
CONE. (male) yellow, small (1/8″); (female) reddish green
- sex – dioecious

fruit . . .
CONE. fleshy, blue (1/8–3/8″), BERRY-like with whitish bloom; cone scales fused
- season – maturity first year

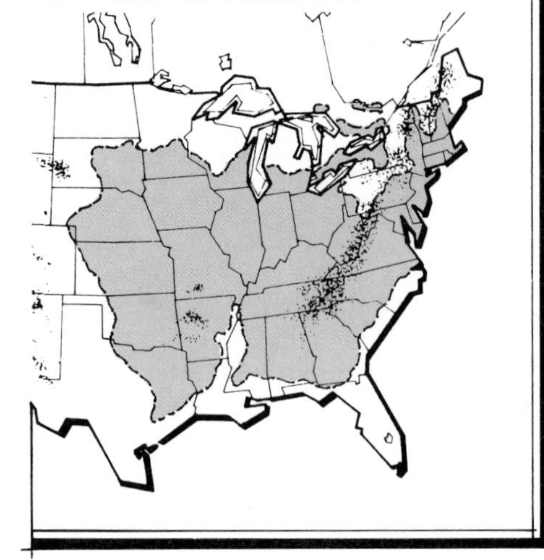

Juniperus virginiana · Eastern Redcedar

Bigtooth Aspen is a medium-sized, upright tree that can be attractive as a specimen plant but is much more at home in a cluster planting. It is a tree of the northern United States and Canada that is very intolerant of shade and flooding. Bigtooth Aspen usually grows on dry sunny slopes near the tops of hills. It likes to grow near Quaking Aspen or pine stands in wooded openings that have been cleared by fire, disease, or some other forest disturbance.

Bigtooth Aspen is a member of the Poplar Family, and it may be properly identified by the large teeth at the leaf margin and by the white, cottony covering on the lower surface of the leaves when the tree is a young sapling.

Aspens are fast-growing trees that, like sumac, recover quickly after a fire by sending up new growth from the roots. Bigtooth Aspen, along with Bur Oak and other members of the Savanna Community, battle the prairie for the right to occupy the land.

In the fall, the whitened bark and yellow leaves are two obvious reasons for aspen's popularity with people who are viewing or photographing the landscape.

habitat . . .
SAVANNA. upland mesic, mesic-dry, lower slopes, terraces and upland ridges
- zone – 3a

form . . .
COLUMNAR. narrow, small canopy tree (50–75′)
- branching – PICTURESQUE with slender, clear trunk
- twig – stout, gray-brown, hairy (young) with clustered, sharp-pointed end buds; gray, WOOLLY scales
- bark – SMOOTH, greenish orange becoming dark gray-brown; (old) trunks deeply furrowed at base

foliage . . .
ALTERNATE. simple, ovate to oblong-ovate, dull, bright green leaves (2–3½″) with large, coarse-ANGLED tooth margin; long, FLATTENED petiole, white, WOOLLY beneath (young)
- color (fall) golden YELLOW
- season – deciduous

flower . . .
CATKIN. slender, silvery-gray (1–2½″), WOOLLY clusters; pendulous
- sex – dioecious, appearing before leaves

fruit . . .
CAPSULE. small, yellow-green (¼″) dia., CATKIN-like clusters (4″) with COTTONY seeds; drooping
- season – maturing early spring after leaves

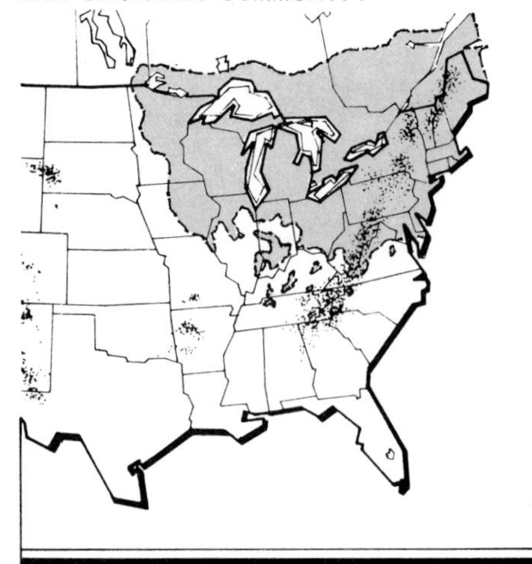

Populus grandidentata · *Bigtooth Aspen*

Quaking Aspen is very widely distributed throughout the northern United States, Canada, and the western mountain regions. It is often a companion tree to Bigtooth Aspen and grows in similar places, such as dry sunny slopes near the tops of hills.

Quaking Aspen is also in the Poplar Family and is easily identified by its small leaves, which are nearly round except for the pointed tip. The leaf petiole has a flattened surface that allows the shiny leaf to quiver in the slightest breeze. I like to think of Quaking Aspen as a happy tree, for it always appears to be laughing when its leaves rattle together. The quaking or trembling character suggested by this plant's name seems somehow inappropriate to its nature.

The tree reproduces rapidly from both seed and root suckers and grows in abundance in fire-damaged and cutover areas. It is a short-lived tree of fifty to sixty years, like the Bigtooth Aspen.

Pulp from aspen is used to make paper for books and magazines. Having no odor, its wood is used for crating cheese and other foods (Schoonover, 1958).

Populus tremuloides · Quaking Aspen

habitat . . .
SAVANNA. upland mesic, mesic-dry, rocky streams, north and east facing slopes
- zone – 2

form . . .
COLUMNAR. narrow, large understory tree (35–50′)
- branching – LEGGY, picturesque with stout, clear trunk
- twig – slender, gray-brown, hairless with smooth, glossy, clustered, SHARP-pointed end buds; red-brown
- bark – SMOOTH, white to green with creamy, WAXY appearance having brown-black WARTY patches; deep-furrowed base

foliage . . .
ALTERNATE. simple, CORDATE, sharp-tipped, bright green leaves (1½–3″) with fine, rounded, SAW-toothed margin; petiole long, flattened, RUSTLING among slightest of breezes
- color (fall) bright YELLOW
- season – deciduous

flower . . .
CATKIN. slender, silvery-gray (1½–4″), FUZZY clusters; pendulous
- sex – dioecious, appearing early spring before leaves

fruit . . .
CAPSULE. slender, yellow-green, (¼″) dia., CATKIN-like clusters (4″) with COTTONY seeds; drooping
- season – maturing early spring

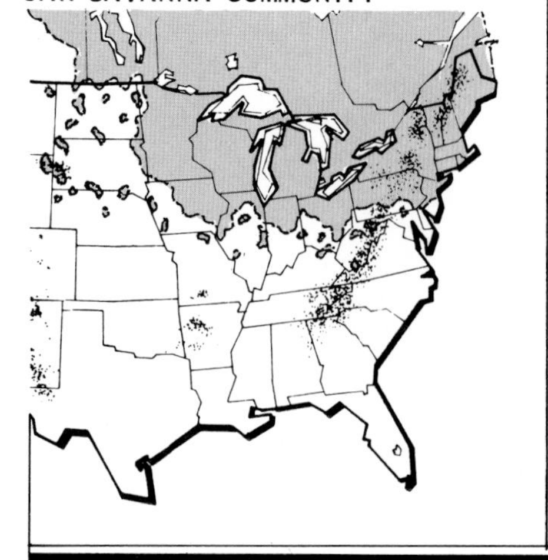

Northern Pin Oak has only been known to science since the 1920s when it was discovered and named by J. E. Hill, a student of oaks (Grimm, 1983). Northern Pin Oak, or Hill's Oak, is very similar to Pin Oak. It has the same general appearance, with its abundance of spur-like branches. The leaves are very similar, deeply divided into five to seven bristle-tipped lobes.

When one examines the acorn, however, the similarity ends. Northern Pin Oak has acorns that are slender and elliptical in shape while Pin Oak acorns are small and button-like with a thin saucer cap. The detection of the two distinct types of acorns led to the further observation that the producer of elliptical acorns also has a more northerly range and finds its home on drier and sandier sites.

The Northern Pin Oak pictured stands very near the water at Pine Lake State Park in Eldora, Iowa. The tree has been there much longer than the artificial lake. Unfortunately, judging from the angle of the tree bent over the water, the lake will be there longer than the tree.

Northern Pin Oak is a slow grower, long-lived, and difficult to transplant. Therefore, it is not readily available at nurseries.

habitat . . .
SAVANNA. upland dry, gravel and sandy upland hills
- zone – 3b

form . . .
OVOID. pyramidal, small canopy tree (50–75′)
- branching – ASCENDING crown with short, descending limbs
- twig – slender, red-BROWN, short, PIN-like with clustered, sharp-pointed, red end buds; scales hairy
- bark – smooth, gray-brown, shallow-fissured with KNOBBY plates

foliage . . .
ALTERNATE. simple, pinnately LOBED, bright green leaves (3–5″) having OPEN sinuses with bristle tip; glossy
- color (fall) SCARLET-red
- season – deciduous, PERSISTENT through mid-winter

flower . . .
CATKIN. slender, YELLOW-green (1½–2″) open clusters; pendent
- sex – monoecious

fruit . . .
ACORN. small, chestnut brown (½–⅝″), ELLIPTIC, ½ enclosed by deep, BOWL-shaped cup with gray, hairy scales
- season – maturing second year

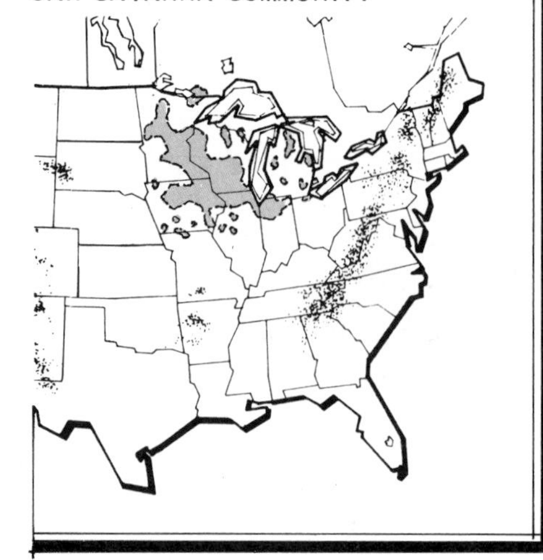

Quercus ellipsoidalis · Northern Pin Oak

Bur Oak, or Mossycup Oak, is the Midwest's greatest oak. It leads the battle against prairie grasses for possession of the land. The deeply furrowed bark that covers even the twigs of seedling Bur Oaks is fire-resistant and enables the tree to survive prairie fires, establishing a foothold where other forest trees can eventually follow. Bur Oak is a very adaptable tree. It is found on sites ranging from dry, south- and west-facing slopes and prairie margins to wet floodplain terraces. It is a very stout-trunked, coarsely branched tree with a broadly rounded crown. Bur Oak is usually wider than it is tall; it often grows 75 feet in height and spreads 100 feet in width. The leaves are thick with a glossy, dark green color above and silvery white beneath, giving the tree a sparkling appearance from a distance when the wind blows the foliage. The typical Bur Oak leaf appears to be cut almost in two where two opposite sinuses try to meet. The acorn has an extradeep cup, covering at least half the nut. The cup's outer fringe suggests a small bird's nest. The wood of the Bur Oak is similar in appearance to White Oak, but Bur Oak is much superior in strength. Bur Oak grows slowly and lives for 200 to 300 years.

habitat . . .
FOREST, SAVANNA. lowland wet-mesic, upland mesic-dry, dry, floodplains, south and west facing slopes
- zone – 2

form . . .
GLOBULAR. stout, open, large canopy tree (75–100′)
- branching – massive, with broad, SPREADING horizontal limbs
- twig – stout, red-brown, winged with CORKY ridges having clustered, tan-brown end buds; blunt
- bark – GRAY-brown, deep-furrowed with loose, RECTANGULAR plates

foliage . . .
ALTERNATE. simple, pinnately LOBED, oblong-obovate, dark green leaves (2–5″) with shallow, ROUNDED lobes; center sinuses reaching midrib; glossy above
- color (fall) yellow-BROWN
- season – deciduous

flower . . .
CATKIN. slender, YELLOW-green (1–2″), open clusters; drooping
- sex – monoecious

fruit . . .
ACORN. large, tan-brown (¾–1¼″), ELLIPTICAL, ½ enclosed by warty, BURLY, fringed cup; gray scales
- season – maturing first year

OAK SAVANNA COMMUNITY

Quercus macrocarpa · Bur Oak

Downy Hawthorn is a small tree, reaching heights of 35 to 50 feet. It is common along fencerows, abandoned fields, and the edges of woods and pastures, often in association with Gray Dogwood and Smooth Sumac. It prefers sunny locations and moist soils but tolerates droughty conditions as well.

The white, five-petaled, rose-like blossoms appear with the leaves from late May through early June and are followed by tiny, red apples that hang on into winter. Roses, apples, and hawthorns all belong to the same family.

The leaves are as wide as they are long, with many large teeth or lobes that in turn contain many smaller teeth. This characteristic is referred to as doubly serrate or twice-toothed.

This slow-growing tree, with its tough wood, thorns, and rose-like flowers, is a desirable tree. Safe from cats, hawks, and children, birds find Downy Hawthorn's thorny branches to be veritable citadels in which to build their nests. Goldfinches, cardinals, Cedar Waxwings, and robins all find refuge among its branches.

Hawthorns in general are affected by rust blights and insect damage. Downy Hawthorn is especially susceptible and should not be planted near Eastern Redcedar, which is the alternate host of several rusts.

Crataegus mollis · Downy Hawthorn

habitat . . .
FOREST, SAVANNA. upland dry and lowland wet-mesic, rocky hillsides, streams and woods edge
- zone – 3b

form . . .
GLOBULAR. broad, large understory tree (35–50′)
- branching – dense, HORIZONTAL, wide-spreading limbs
- twig – slender, gray-brown; FLAKY with glossy brown (2″) thorns having reddish, DOME-shaped buds
- bark – SHAGGY, silvery gray, twisting scaly plates

foliage . . .
ALTERNATE. simple, oblong-ovate LOBED, yellow-green leaves (3–4″) with DOUBLY saw-toothed margins; tomentose, dull above with hairy stem
- color (fall) yellow-BROWN
- season – deciduous

flower . . .
CLUSTER. small, white (¾″), open blooms, FLAT-TOPPED clusters often 3–4″ across
- sex – monoecious

fruit . . .
BERRY. dull RED, (¾–1″) dia., APPLE-like drooping clusters
- season – maturing in late summer

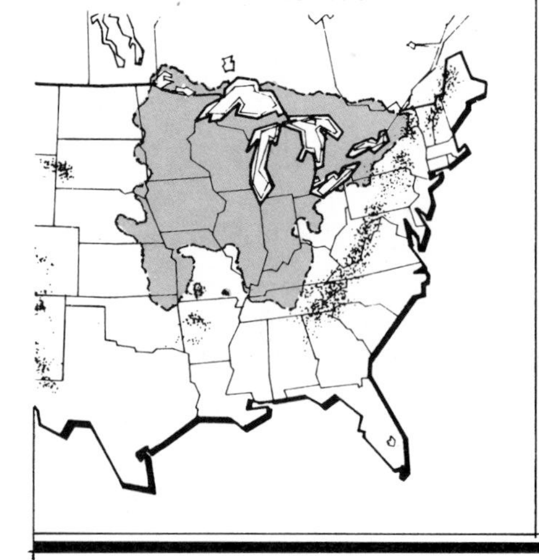

Prairie Crabapple, or Iowa Crabapple, is a small tree that grows nearly as wide as it does tall (25 feet). It is commonly found throughout the central Midwest on dry upland soils of open woods, pastures, and rocky hillsides.

Like so many of the woods' edge trees and shrubs, Prairie Crabapple has very coarse, stiff branches and twigs with modified thorns to protect it from large browsing animals.

The pink blooms that appear in late May and early June have been considered by many as the most beautiful and fragrant of the native wild crabapples. In fact, the Bechtel Crabapple, with its large, double, rose-colored blossoms was developed as a cultivar from this species (Grimm, 1983).

The fruit is an apple that is 1 inch in diameter. It matures to a yellow-green color and is an important food source for songbirds and small mammals.

habitat . . .
SAVANNA. upland dry, rocky hillsides, creek banks and woods edge
- zone–2

form . . .
GLOBULAR. open, small understory tree (20–35′)
- branching–rigid, HORIZONTAL to irregular limbs
- twig–slender, brown; gray flaky skin with small, torn-like, right-angle SPURS having bright red buds ; blunt
- bark–SCALY, brown, silvery gray, exfoliating PAPERY sheets

foliage . . .
ALTERNATE. simple, oblong, shiny bright green leaves (2½–4″) with double-toothed, somewhat 3-LOBED margin; base wedge-shaped, hairy underside with prominent veins; tomentose
- color (fall) YELLOW-orange
- season–deciduous

flower . . .
SINGLE. large, PINKISH white, (½″) dia., clusters in groups of 3–6 on long stems
- sex–monoecious, early spring, FRAGRANT

fruit . . .
BERRY. single, yellow-green, (1–1½″) dia., APPLE-like, long-stalked; pendent
- season–maturing late summer

OAK SAVANNA COMMUNITY

Malus ioensis · Prairie Crabapple

American Plum is a common small tree found in dry upland pastures, fencerows, woods' edges, thickets, and stream banks. It has also been used throughout the Midwest in windbreaks. It occasionally grows 20 to 35 feet tall but is typically shorter. It grows wider than it is tall.

The branches of American Plum have horizontal lenticels. The bark on the trunk tends to peel and exfoliate. Like many plants of the savanna, American Plum has thorn-like twigs to discourage browsers.

In spring, American Plum is one of the first trees to bloom, with flowers opening just before the leaves. The flowers are very fragrant and attract many bees and other insects. The plant's main attraction, however, is its fruit. The sugary, fleshy, 1-inch-diameter fruit is produced in great abundance, much to the delight of people and animals. Many birds are attracted to the red plums, and people eat them freshly picked or use them in preserves and wine.

American Plum is a fast grower and has been used for erosion control, but it is short-lived, surviving only thirty-five to sixty-five years.

habitat . . .
SAVANNA. upland dry, woods edge, steep rocky hillsides, stream banks and thickets
- zone – 3a

form . . .
GLOBULAR. broad crown, small understory tree (20–35′)
- branching – stiff, horizontal, SPREADING limbs with short, blue-gray spurs
- twigs – slender, red-brown, FLAKY skin often ending in short spine with clustered, sharp-pointed end buds; brown
- bark – dark brown, TWISTING, exfoliating, red-tinged to CURLING along margin with horizontal lenticles

foliage . . .
ALTERNATE. simple, elliptic-oval dark green leaves (2½–4″) with toothed margin; deep-seated veination, slightly tomentose below with glands on petiole; long-pointed tip
- color (fall) golden YELLOW
- season – deciduous

flower . . .
CLUSTER. small, white with red tinge, (½″) dia., FLAT-TOPPED cluster in groups of (2–5) on slender stalk; fragrant, SPICY
- sex – monoecious, appearing before leaves

fruit . . .
BERRY. fleshy, RED to purple, (1–1½″) dia., PLUM-like in groups of 2–3
- season – maturing early summer

Prunus americana · American Plum

Common Chokecherry is a small tree or large shrub that is tolerant of many soils and locations. It is found in thickets along roadsides and streams, at the edges of woodlands, and scattered among hardwoods. It is one of the most widely distributed trees in North America, as it is hardy, fast growing, and moderately long-lived.

Common Chokecherry has attractive white flowers that are quite similar to those displayed by Black Cherry. The fruit, though eaten by people and animals, should be avoided until it is fully ripened and loses its astringent, "puckery" quality. Jellies and pies are made from the fruit. The Eastern Bluebird is one of many bird species that uses the fruit of this native plant as a food source (Degraaf and Whitman, 1979).

The bark of Common Chokecherry is covered with vertical lenticels, giving the appearance of stretch marks from rapid growth. The twigs, which are very pungent when broken, develop cankers along the stems from insect stings.

habitat . . .
FOREST, SAVANNA. upland mesic-dry, open slopes, woods edge and open fields
- zone – 2

form . . .
OVOID. irregular, large understory tree (35–50′)
- branching – stout, horizontal, ASCENDING limbs
- twig – slender, red-brown, (BITTER) aromatic with brown clustered buds; margin scales white
- bark – smooth, ash gray with CONSPICUOUS lenticles

foliage . . .
ALTERNATE. simple, oblong-lanceolate, shiny bright green leaves (1½–3″) with sharp, SAW-toothed margin; slightly thickened, (2) small glands on petiole
- color (fall) yellow-ORANGE
- season – deciduous

flower . . .
SPIKE. small, white, (½″) dia., dense, pyramidal (3–6″) clusters; pendulous
- sex – monoecious, late spring, FRAGRANT

fruit . . .
BERRY. small, PURPLE-black, (¼–⅜″) dia., open (2″) clusters; pendent
- season – maturing in summer

OAK SAVANNA COMMUNITY

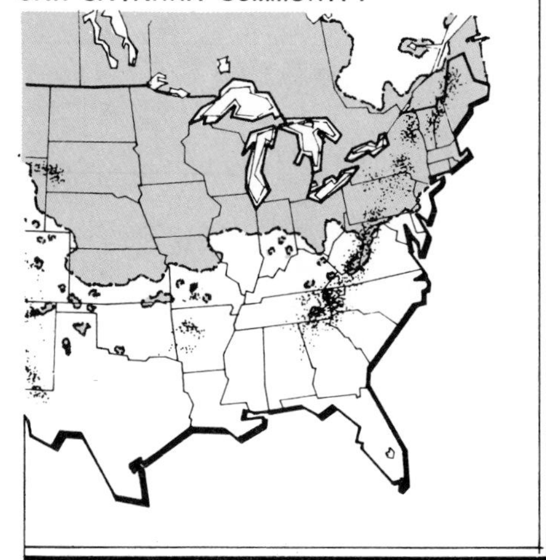

Prunus virginiana · Common Chokecherry

Blackhaw Viburnum is one of a large group of attractive plants related to the elders and belongs to the Honeysuckle Family. Viburnums generally share many favorable qualities. All have clusters of flowers, usually white, that are both pretty and fragrant. They all produce colorful fruit, and the fall leaf coloration of most viburnums is an appealing deep maroon.

Blackhaw Viburnum shares all of these qualities. It is often crooked-stemmed and rather scraggly when young, but it frequently matures into a sturdy and shapely small tree with opposite spine-like branches.

The shape and character of its leaves are plum-like. The fruit is a dark blue, oblong berry held in loose clusters on red stems. The fruit, like that of Nannyberry Viburnum, has a thin, sugary flesh that is edible. This fruit, with its very large seeds, is eaten by a variety of birds and mammals, and it is widely distributed by birds.

Blackhaw Viburnum is generally found on dry, well-drained soils, often forming thickets along fencerows and bordering the edges of woodlands.

Blackhaw Viburnum is an excellent ornamental shrub. Unlike most other viburnums, it is intolerant of shade and should be given a sunny location.

habitat . . .
SAVANNA. upland dry, steep rocky hillsides, creek and stream banks and woods edge
- zone – 3b

form . . .
OBOVOID. dense, small understory tree (20–35′)
- branching – stiff, TWIGGY, horizontal spreading limbs; spur-like angled branchlets
- twig – smooth, slender, gray-brown with naked FLESH-colored, sharp-pointed end buds; SWOLLEN base, glabrous
- bark – rough, brown-black with small, SQUARISH plate

foliage . . .
OPPOSITE. simple, elliptic-oval, bright green leaves (1½–3″) with fine-toothed margin; blunt apex with slightly rounded base, dull beneath, having prominent veination; petal often with narrow margin
- color (fall) scarlet-RED
- season – deciduous

flower . . .
CLUSTER. small, white, (¼″) dia., open, FLAT-TOPPED clusters (3–4″) across; slightly PROLIFIC
- sex – monoecious, early spring, fragrant

fruit . . .
BERRY. small, blue-black, (½″) dia., FOOTBALL-shaped, drooping clusters on reddish STEMS; terminal
- season – maturing autumn, PERSISTING through mid-winter

OAK SAVANNA COMMUNITY

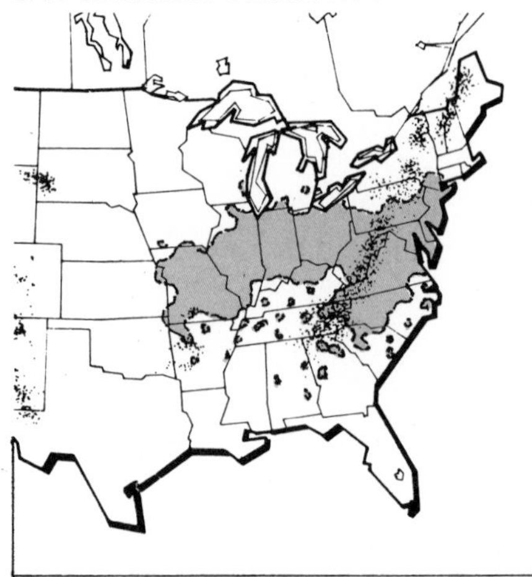

Viburnum prunifolium · Blackhaw Viburnum

Gray Dogwood is one of the most common native shrubs. It is an excellent choice for both landscaping purposes and wildlife habitat. It grows on a variety of sites from dry uplands and woods' edges to moist, rich lowlands. Gray Dogwood adapts to a wide range of soils. Tolerating some shade, it is often abundant along roadsides, in field fencerows, and in prairie thickets.

In early June, Gray Dogwood is in full bloom with small clusters of creamy white flowers. By September, the flowers develop into showy clusters of white, berry-like fruit on red stalks. The fruit if not eaten by hungry birds, contrasts beautifully with the leaves, which deepen to wine-red or purple in the fall.

If planted in a natural setting, Gray Dogwood blends handsomely with either Smooth Sumac on hills or Common Ninebark in valleys.

habitat . . .
FOREST, SAVANNA. mesic-dry, dry, wet-mesic alluvial woods, woods edge, rocky ledges and hilltops
- zone – 3a

form . . .
OBOVOID. open, large-medium shrub (6–15′)
- branching – erect, ARCHING with stout branchlets
- twig – smooth, gray-brown with whitish brown pith and small buds; reddish brown
- bark – gray, smooth with rough lenticles

foliage . . .
OPPOSITE. simple, elliptic-lanceolate, dark green narrow leaves (2–4″) with entire margin; veination, (3–5) pairs parallel to leaf edge
- color (fall) reddish PURPLE
- season – deciduous

flower . . .
CLUSTER. small, white, 2″ across, ROUND-topped clusters; profuse
- sex – monoecious, FRAGRANT

fruit . . .
BERRY. small, white, (3/8″) dia. clusters (4″) across; supported on RED stems
- season – maturing late summer

Cornus racemosa · Gray Dogwood

American Filbert, or Hazelnut, grows in dense clumps or thickets and along fencerows on a variety of soils. In naturalistic plantings, it is gaining popularity as an ornamental because of its interesting catkins, nuts, and autumn leaf color. The catkins hang stiffly throughout the winter but relax as they develop in the warmth of spring, attracting early bees that spend long hours gathering the pollen. The pistillate flowers on the same plant are little star-like tufts of crimson.

The fruit of American Filbert is a small, brown nut that resembles the commercial variety in color and flavor. In fact, some consider it superior. The leafy husk covering the nut, which is difficult to see until fall, turns brown and dries to reveal the nut within. This fuzzy, hazelnut fruit is popular with many birds and mammals as a food source.

The leaves of the American Filbert are doubly toothed and fuzzy on the underside, giving the leaf a soft, velvety texture. The fall leaf color is quite attractive and varies from bright yellow to deep wine-red. As the leaves fall in autumn, the nuts are clearly visible—if squirrels have not already taken them.

habitat . . .
SAVANNA. upland mesic, mesic-dry, alluvial woods, stream banks and rocky open hillsides
- zone—3a

form . . .
GLOBULAR. open, medium shrub (6–15′)
- branching—upright, ASCENDING stems, often forming spreading clumps
- twigs—stiff, gray-brown; hairy with blunt, oval buds; false end buds
- bark—gray-brown, smooth, often peeling

foliage . . .
ALTERNATE. simple, obovate-elliptic, yellow-brown leaves (2–5″) with HEART-shaped base, coarse-toothed margin; hairy above, downy beneath; stalks tomentose
- color (fall) reddish MAROON
- season—deciduous

flower . . .
CATKIN. (male) gray-brown (1½–3″), slender, PENCIL-like, pendulous cluster; (female) RED (1/16–1/4″), erect
- sex—monoecious, early spring

fruit . . .
NUT. reddish brown, (½″) dia., encased by thin, RAGGED-edged papery husk; flattened in groups of 2s
- season—maturing late autumn

OAK SAVANNA COMMUNITY

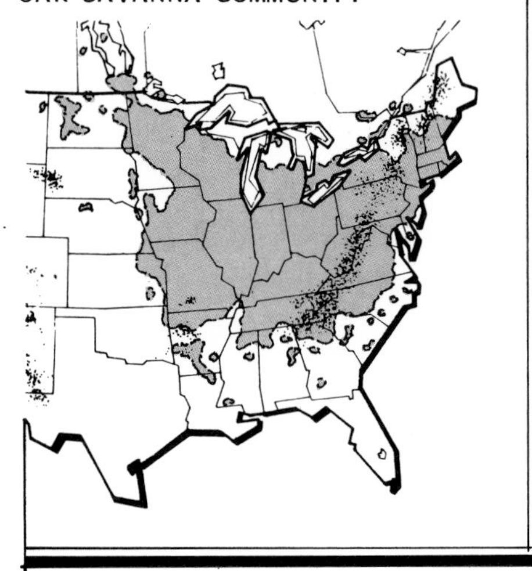

Corylus americana · American Filbert

Smooth Sumac is a prairie relict that frequents roadsides, thickets, fencerows, and pastures. It grows from underground rhizomes that were once its defense against prairie fires. Smooth Sumac is a hardy, rapid-growing shrub that needs very little care and prefers sandy soils and gravelly upland sites. However, it is intolerant of shade.

Its name, Smooth Sumac, comes from the smoothness of its twigs and leaf petioles as compared with Staghorn Sumac, which has a velvet fuzz covering its new growth.

The bloom of Smooth Sumac is an upright cluster of greenish-yellow flowers that matures in the fall to a pyramidal cluster of velvety red berries. These clusters remain on the shrub, changing slightly in color, throughout the winter. Sumac's compound leaves are among the earliest to change color in autumn, becoming brilliant scarlet and crimson, brightened with yellow and orange.

Smooth Sumac has become increasingly popular in urban planting schemes because of its fast growth and brilliant fall color. It is most effective, however, when planted in drifts, or colonies, that are typical of natural settings.

habitat . . .
SAVANNA. upland dry, steep rocky bluffs and open woods
- zone—2

form . . .
IRREGULAR. obovoid, medium shrub (6–15′)
- branching—coarse, crooked, LEGGY branchlets
- twig—stout, red-purple, hairless, somewhat bluish BLUME with leaf scar encircling small, DOME-shaped bud; velvety white
- bark—smooth gray becoming scaly with conspicuous DOT-like lenticles

foliage . . .
ALTERNATE. pinnately COMPOUND, leaflets (9–27) oblong-lanceolate, glossy dark green leaves (2–5″) with coarsely toothed margin; whitish bloom beneath
- color (fall) SCARLET-red
- season—deciduous

flower . . .
CLUSTER. (male) brown, erect; (female) yellow, small, (1/8″) dia., dense (3–5″) pyramidal clusters
- sex—dioecious

fruit . . .
BERRY. small, dark RED, (1/8″) dia., dense, erect (3–6″) pyramidal clusters with short, STICKY, crimson hairs; velvety
- season—maturing late summer, PERSISTING through winter

OAK SAVANNA COMMUNITY

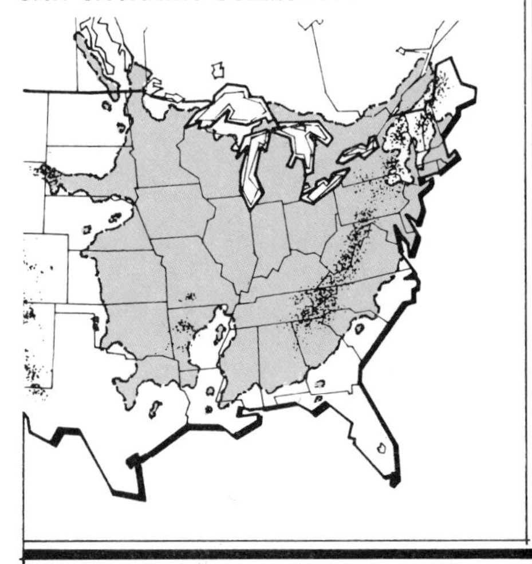

Rhus glabra · Smooth Sumac

Missouri Gooseberry is a widespread shrub quite common throughout the Midwest. It grows in open wooded pastures and deep woods and in either rich or rocky soil. Like most other gooseberries, it is a small, thorny plant armed with three nodal spines and scattered, bristle-like prickles on the lower branches.

Missouri Gooseberry fruit, however, is entirely smooth. It ripens gradually from late June to September. As it matures, it is tart and yet quite pleasing in taste. Gooseberry fruit is eaten by many birds and mammals and is used by people in jelly, preserves, and pies. The shrub, with its armed spines, also provides excellent cover for wildlife.

habitat . . .
SAVANNA. mesic-dry, dry, wet-mesic, alluvial woods, woods edge and rocky wooded stream banks
- zone–4a

form . . .
OBOVOID. upright small shrub (4–6′)
- branching–stout, dense, ARCHING limbs
- twigs–smooth, dark red-brown to brown with infrequent spines having small, spreading scale buds; red-brown
- bark–exfoliating into thin, lateral PAPERY curls, grayish red-brown; ARMED with stout spines

foliage . . .
ALTERNATE. simple, cordate, yellow-green leaves (2–4″) with 3–5 LOBED margin; rounded lobes; pubescent beneath
- color (fall) yellow-ORANGE to red, then purple
- season–deciduous

flower . . .
TUBULAR. slender, whitish green (1/8–1/2″), TRUMPET-shaped with recurved petals at tip in groups of 2 to 3; PENDULOUS
- sex–dioecious

fruit . . .
BERRY. small, purplish black, (1/8–1/4″) dia., smooth; TART
- season–maturing late summer

OAK SAVANNA COMMUNITY

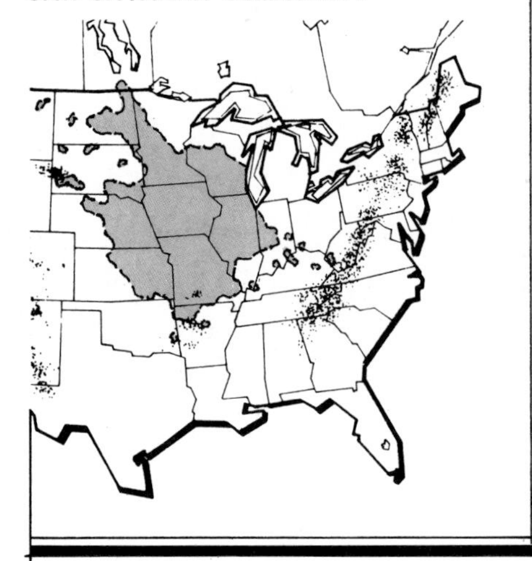

Ribes missouriense · Missouri Gooseberry

Western Snowberry grows in large patches or drifts of plants in open areas, often on gravelly hillsides or along ravines. It is an excellent shrub for controlling soil erosion and offers protection for wildlife.

The smooth, round fruit of Western Snowberry is not eaten by birds or mammals. It remains attached, then withers and falls from the plant in the middle of winter. The fruit tends to be greenish-white and is not as white as White Snowberry *(Symphoricarpos alba)* but is still very attractive and heavily clustered, causing the branches to arch to the ground.

The small, bell-shaped, pink flowers are produced almost continuously from June until September. The leaves of Western Snowberry are opposite, as are those of all *Symphoricarpos,* but they tend to be larger than others in the family.

Other common names are Snowberry, Wolfberry, and White Buckbrush.

habitat . . .
PRAIRIE, SAVANNA. mesic-dry, dry, open rocky hillsides, ravine banks, open sandy flats and open woods edge
- zone – 2

form . . .
OBOVOID. open, small shrub (4–6′)
- branching – upright, often leggy with ARCHING limbs
- twig – SLENDER, brown, slightly hairy
- bark – gray-brown, thin

foliage . . .
OPPOSITE. simple, ovate-obovate, bright grayish green leaves (2–4″) with wavy margin; leathery, tomentose beneath
- color (fall) GRAY-green
- season – deciduous

flower . . .
CLUSTER. small, PINKISH white, (1/8″) dia., clustered along stem in leaf axils
- sex – monoecious

fruit . . .
BERRY. small, greenish WHITE, (1/16–1/8″) dia. dense clusters along stem
- season – maturing late summer, PERSISTING throughout winter

OAK SAVANNA COMMUNITY

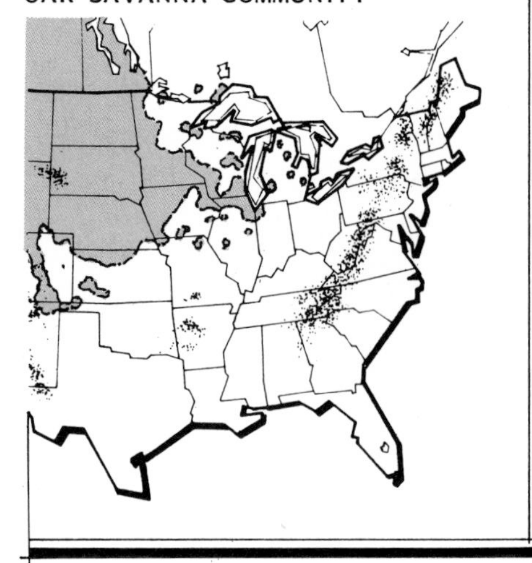

Symphoricarpos occidentalis · *Western Snowberry*

Indiancurrant Coralberry, or Buckbrush, grows in a wide variety of dry and moist soils. It tolerates both sun and shade and is commonly found on dry, rocky banks and in open woods, where it tends to form dense thickets. The flower nectar attracts ruby-throated hummingbirds (Degraaf and Whitman, 1979), and the red, clustered fruit is a favorite food of many game- and songbirds. The abundance of fruit on the arching stems of the Coralberry is little short of amazing. The fruits clustered in the axils of the opposite leaves are so crowded that they completely surround the stem. One stem 7 inches long was found to hold 250 currants (Keeler, 1903). Indiancurrant Coralberry is able to succeed on dry, barren, and inhospitable sites. As a result, it is an excellent shrub to use for controlling gullies and soil erosion along road banks.

It is also a good shrub for borders in garden plantings. The brightly colored fruit is attractive with or without the foliage, which frequently persists on the branches well into the winter months.

habitat . . .
SAVANNA. upland mesic-dry, upland dry, wet-mesic, floodplain, woods edge, stream banks, rocky hillsides and open roadsides
- zone – 2

form . . .
UPRIGHT. open, leggy, small shrub (4–6′)
- branching – stout, ARCHING, spreading branchlets; slightly hairy
- twig – smooth, brown, solid white pith with small CINNAMON scaly buds; present at branch axils
- bark – thin, PAPERY, brown

foliage . . .
OPPOSITE. simple, elliptic-ovate, yellow-green leaves (3/4–1 1/2″) with somewhat WAVY margin; short-stalked, base WEDGE-shaped, tomentose beneath
- color (fall) yellow-GREEN
- season – deciduous

flower . . .
CLUSTER. small, yellowish white (1/8–1/4″), TRUMPET-shaped, dense, axillary clusters or terminal spikes; pinkish TINGED
- sex – monoecious

fruit . . .
BERRY. small, reddish PURPLE, (3/8″) dia., dense clusters along stems in leaf axils
- season – maturing late summer, PERSISTING through autumn

OAK SAVANNA COMMUNITY

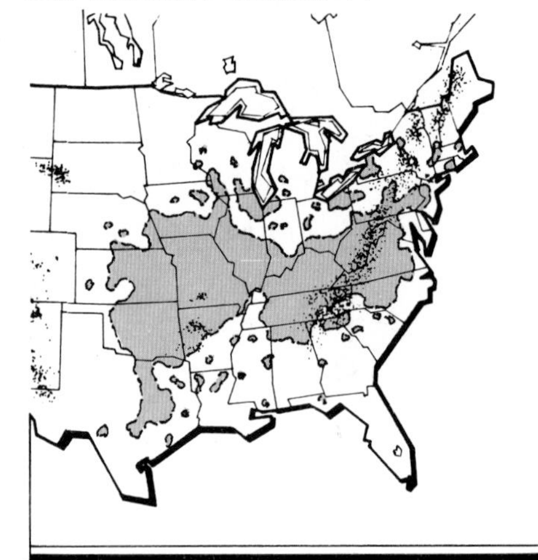

Symphoricarpos orbiculatus · Indiancurrant Coralberry

Oak-Hickory Community

The Oak-Hickory community is found predominantly on the south- and west-facing slopes in the midwestern forest region. In this region, the Oak-Hickory community is characteristically bounded by bluestem prairie or oak Savanna on the top of the slope and by the Floodplain community on the bottom of the slope.

Exposure to wind and direct summer sun causes surface temperatures under an Oak-Hickory canopy to rise 8 to 10 degrees more than under a Maple–Linden canopy (Aikman, 1948). Soil moisture content, therefore, sometimes drops to dangerously low levels. Therefore, plants that cannot adjust to high temperatures and droughty soils are not found in this community.

The Oak-Hickory community has a greater number of dominant canopy tree species than does the Maple–Linden community, but its canopy is less dense, allowing more sunlight to reach the shrub layer. The resulting diversity of the shrub layer in the Oak-Hickory community is one of its conspicuous features.

The most common understory tree of the Oak-Hickory community is American Hophornbeam, which retains most of its leaves throughout the winter. During this season, when the canopy trees have lost most of their leaves, American Hophornbeam can be easily identified by the many small crowns of light brown, curled leaves that dot the sunny slopes.

Trees: Canopy (over 48′)

Trees: Understory (12–48′)

Shrubs: (6–12′)

Shrubs: (less than 6′)

Vines:

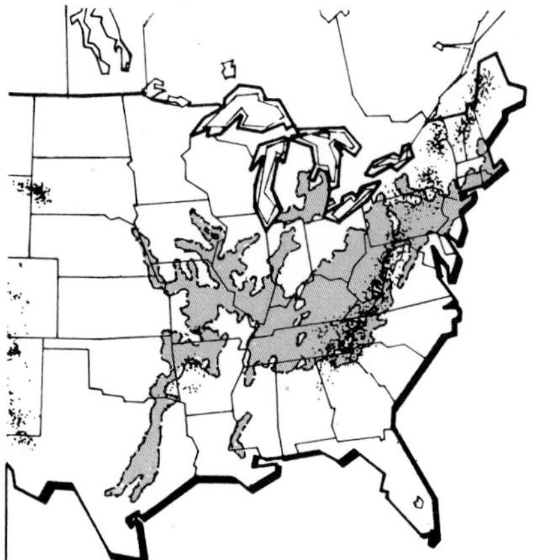

Shagbark Hickory has an important place in our pioneer heritage. The strength and straightness of this tall, enduring tree made the nickname, Old Hickory, a natural for our seventh president, Andrew Jackson. In pioneer times, Shagbark Hickory contributed tough, durable hoe handles and whiffletrees for farm wagons. It gave bushels of sweet-kerneled nuts and cords of perfect firewood. Hickory was also the primary fuel for the smokehouse curing of meats.

Shagbark Hickory grows in dry, upland woods, along with its respected companions, Bur Oak and White Oak. The smooth, dark gray bark of a young Shagbark Hickory becomes increasingly rough as it matures, until the long, partially detached strips of an adult tree's bark make it easy to recognize. The exfoliating bark and straight trunk make Shagbark Hickory a handsome yard tree.

Shagbark Hickory grows at a moderate rate and is long-lived, surviving 200 to 300 years. The tree's large taproot makes transplanting difficult after the seedling stage.

habitat . . .
FOREST, SAVANNA. upland mesic-dry, alluvial valleys, south and west facing slopes
- zone—4a

form . . .
OBOVOID. irregular, large canopy tree (75–100′)
- branching—stout, ASCENDING limbs with lower branches drooping
- twig—stout, reddish to gray-brown, ending in blunt, egg-shaped, brown PAPERY buds; EXFOLIATING scales, hairy
- bark—smooth, slate gray, separating into VERTICAL exfoliating, flattened, SHAGGY plates

foliage . . .
ALTERNATE. pinnately COMPOUND, leaflets (5–7), oblanceolate, yellow-green leaves (3–7″) with fine, saw-toothed, hairy margin; base wedge-shaped, stalkless with long-pointed apex
- color (fall) YELLOW-brown
- season—deciduous

flower . . .
CATKIN. slender, yellow-GREEN (3–4″), drooping, pendulous clusters in groups of 3; female in pairs at tip of twig
- sex—monoecious, appearing before leaves

fruit . . .
NUT. globular, BROWN becoming black, (1½″) dia., encased by 4-ribbed, THICK husk; nut 4-angled (edible)
- season—maturing early autumn

OAK-HICKORY COMMUNITY

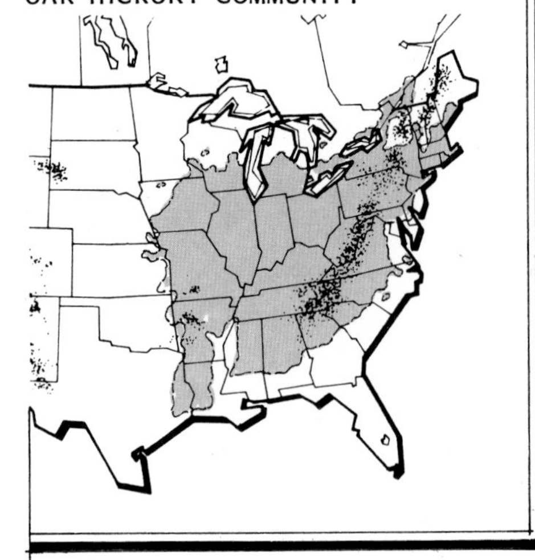

Carya ovata · Shagbark Hickory

Cucumbertree Magnolia is a large, round-topped tree that grows 75 to 100 feet tall and ranks as the hardiest of all magnolias. It is a tree of the Allegheny Mountains but is also found as far west as Missouri and northern Arkansas. It prefers sheltered coves, lower slopes, and protected valleys but grows in any good, moist garden soil.

Its name, Cucumbertree, comes from the cucumber-like appearance of the 3-inch fruit. This curious fruit is first green and then turns dull red. As with all magnolias, the orange-red seeds dangle from the fruit capsules on white thread-like connections when fully ripe.

The leaves are an important ornamental feature of this tree. They range from 6 to 10 inches in length, have a deep green summer tone, and are yellow-brown in the fall. The flowers are not very spectacular, since they are only 2 inches across, have a greenish color, and bloom as the leaves appear in late May and early June.

Cucumbertree Magnolia is a good selection for specimen planting on home grounds. It provides ample shade, has a pleasing appearance in all seasons, and is infected by few pests or diseases. The tree grows at a medium rate and generally matures in 125 years.

habitat . . .
FOREST. upland mesic, bottomland ravines, coves, lower slopes and stream banks
- zone—5a

form . . .
OVOID. globular, large canopy tree (75–100′)
- branching—short, spreading to recurving limbs
- twig—stout, reddish brown, ringed at leaf scars, ENCIRCLING twig with large, WOOLLY, naked, gray-green buds; silvery hairs
- bark—narrow, gray-brown, shallow-furrowed with scaly, longitudinal ridges

foliage . . .
ALTERNATE. simple, elliptic-ovate, yellow-green leaves (5–10″) with entire to slightly WAVY margin; leathery with soft hairs beneath
- color (fall) YELLOW-brown
- season—deciduous

flower . . .
SINGULAR. large, yellow-green (2–3″), BELL-shaped with fleshy, erect petals; TERMINAL
- sex—monoecious

fruit . . .
FOLLICLE. oblong, green to red (2–3″), CUCUMBER-like aggregate on stout stem, splitting; seeds SCARLET that hang on by thin thread
- season—maturing early autumn

OAK-HICKORY COMMUNITY

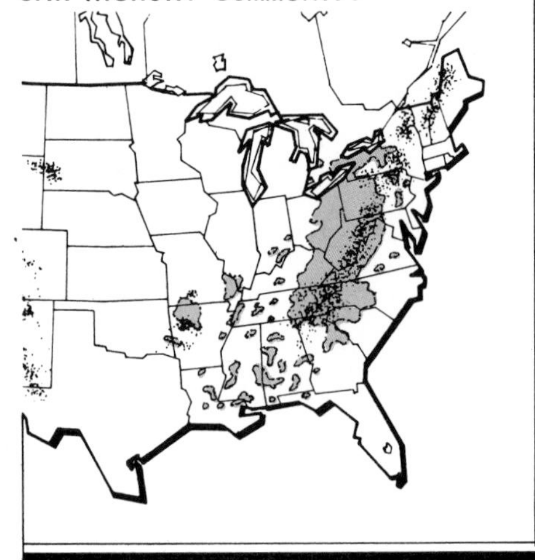

Magnolia acuminata · Cucumbertree Magnolia

Black Cherry is an excellent tree for landscape plantings, but it is seldom used. It is a medium-sized tree, 50 to 60 feet tall, with a narrow, open crown. It is one of the few large trees that has attractive flowers and fruit, the latter of which attracts some seventy species of birds. The tasty fruit can be made into a drink called cherry bounce (Borland, 1983). The fruit has also been used to flavor rum and brandy; in fact, one of the plant's common names was Rum Cherry.

Black Cherry is one of the best hardwoods for high quality lumber. Early American furniture was made from Black Cherry and has passed for mahogany time and time again.

The easiest way to identify Black Cherry is by its distinctive bark. An old trunk is almost black and is made up of small, broken pieces of the original young, smooth cherry bark.

Black Cherry grows on rich, moist soils, commonly mixed with other hardwoods, and it is tolerant of many soil types. Black Cherry is hardy, grows rapidly, and can live 200 years.

habitat . . .
FOREST, SAVANNA. upland mesic and mesic-dry; bottom lands; sandy, rocky slopes, open fields and woods edge
- zone—3b

form . . .
OVOID. columnar, small canopy tree (50–75′)
- branching—stout, ASCENDING with slender, drooping branchlets
- twig—slender, red-brown with conspicuous lenticels having red-brown, LONG-pointed buds, with overlapping scales; leaf scar semicircular
- bark—smooth, thin, red-brown becoming BLACK with horizontal lenticels; fissured to SQUARISH, curving scales

foliage . . .
ALTERNATE. simple, lanceolate, oblong lanceolate, shiny dark green leaves (2–5″) with saw-toothed, incurved margin; thick, leathery, reddish petiole with (2) GLANDS near base of leaf blade
- color (fall) yellow, REDDISH orange
- season—deciduous

flower . . .
SPIKE. perfect, 5-lobed, WHITE (3/8″), pyramidal clusters (4–6″) appearing after leaves; terminal
- sex—monoecious, FRAGRANT

fruit . . .
BERRY. small, red turning black (3/8″), globular, drooping (4–6″) clusters (EDIBLE)
- season—maturing late summer

OAK-HICKORY COMMUNITY

Prunus serotina · Black Cherry

White Oak is a majestic tree valued for its timber as well as its ornamental value. Few trees rate as many compliments. It grows from 50 to 100 feet tall and has nearly perfect proportions, with its short trunk and massive, far-reaching, often nearly horizontal branches. White Oak is partial to dry gravelly soils but grows on a variety of sites.

Its wood is very heavy, hard, and strong. It is close-grained and is one of the best woods for tight cooperage (such as liquor barrels). It is also one of the finest woods for furniture and hardwood flooring and is used for ships, wagons, and railroad ties and as fuel.

White Oak is typical of the White Oak Family in that it produces an acorn crop every year. (Members of the Black or Red Oak family produce acorns in alternate years.) White Oak has whitened bark, edible acorns, buds that are blunt, and leaves that are round-lobed rather than bristle-tipped.

White Oak is the most common of the large woodland trees. It is also the noblest and is king by common consent. It is the finest oak for landscape planting because of its broad-spreading form, its deep green foliage that changes to a magnificent wine-red in autumn, and its abundance of acorns, which are valued as food for wildlife.

habitat . . .
FOREST. upland mesic-dry; well-drained lowlands, upland flats, rocky hillsides, south and west facing slopes
- zone – 4a

form . . .
GLOBULAR. broad crown, large canopy tree (75–100′)
- branching – massive, HORIZONTAL limbs with short, stout trunk
- twig – slender to stout, red-brown with clustered, blunt (3/16″), oval end buds; reddish brown
- bark – scaly, light gray, shallow-furrowed with small narrow ridges

foliage . . .
ALTERNATE. simple, pinnately LOBED, (5–9) oblong, green gray-green leaves (4–9″) with lobes entire having sinuses extending to midrib; whitish, downy beneath
- color (fall) BURGUNDY-red
- season – deciduous, PERSISTING through winter

flower . . .
CATKIN. slender, YELLOW-green (1–2″), singular drooping clusters; staminate
- sex – monoecious

fruit . . .
ACORN. tan-brown, EGG-shaped (3/4–1″), enclosed 1/4 by shallow, WARTY scale cup
- season – maturing first year

OAK-HICKORY COMMUNITY

Quercus alba · White Oak

Scarlet Oak is rated by many as the most beautiful of all the oaks. Its fall color is the richest red of the autumn. In *Some American Trees* (1935), William B. Werthner refers to Scarlet Oak with no small amount of reverence as his "glory tree."

Scarlet Oak is similar to Pin Oak, with leaves deeply cut almost to the midrib. It is distinctly an upland tree that prefers gravelly soils and rocky slopes instead of the low-lying, heavy soils occupied by Pin Oak.

The acorn is the primary means of identification for any oak. The Scarlet Oak acorn is small, sharply pointed, and half-covered by the cup. (Black Oak acorns are similar but have cups covered with loose, hairy scales.) The cup of the Scarlet Oak acorn has tightly pressed scales that appear to be almost sanded smooth.

Growing at a rate of 1½ to 2 feet a year, Scarlet Oak is considered the fastest growing oak. It matures to a medium-sized tree, 70 to 80 feet tall, and lives 200 to 300 years. Trees larger than seedlings, however, are difficult to transplant. Scarlet Oak is a desirable shade tree because of its beauty throughout the seasons.

habitat . . .
FOREST. upland dry, steep rocky ridges, south and west facing slopes
- zone—5a

form . . .
GLOBULAR. open, large canopy tree (75–100′)
- branching—limbs stout with ASCENDING branches
- twig—smooth, RED-BROWN, stout with clustered, gray end buds; WOOLLY scales
- bark—nearly BLACK, smooth plates, becoming furrowed with narrow, scaly ridges; inner bark reddish

foliage . . .
ALTERNATE. simple, pinnately LOBED, glossy bright green leaves (3–7″) with broad "C" sinuses extending to midrib; lobes toothed with bristly tip, silvery white; tomentose below
- color (fall) SCARLET-red
- season—deciduous

flower . . .
CATKIN. slender, YELLOW-green (3–4″), singular drooping clusters; staminate
- sex—monoecious

fruit . . .
ACORN. tan, oval EGG-shaped (½–1″); ⅓–½ enclosed by tan-brown, thick, SMOOTH scaled cup
- season—maturing second year

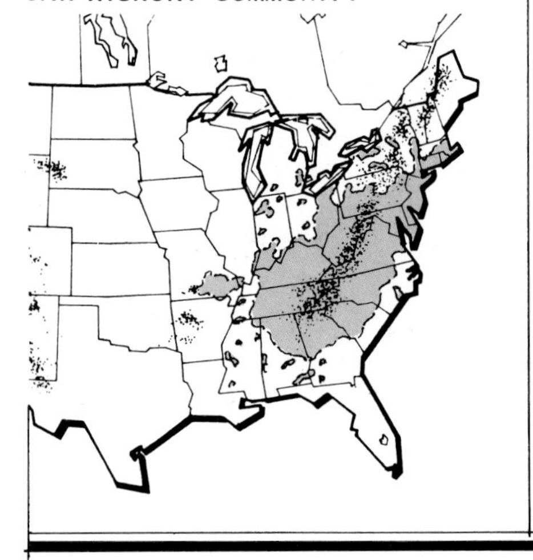

Quercus coccinea · Scarlet Oak

Shingle Oak is a pyramidal tree when young and matures to a tree with a rounded top. It grows on rich hillsides and fertile floodplains.

Both the common name (Shingle Oak) and the scientific name (*imbricaria*, meaning overlapping) attest to the wood's usefulness as shingling material. In pioneer days, many cabin roofs were finished with thin slabs of Shingle Oak wood (Elias, 1980).

On first appearance, the tree does not look like an oak. The leaves are dark green, shiny, and lance-shaped, and they lack the deeply cut sinuses, or rounded lobes, that are typical of most oak foliage. However, several traits confirm that it is indeed a member of the black or red oak group: its acorns mature every other year, its bark is dark colored, and its foliage is bristled, having a sharp point at the tip of each leaf.

The acorns are similar to those of Pin Oak, quite small and button-like, and they are important wildlife food. Turkeys, deer, squirrels, quail, and ducks feed on Shingle Oak acorns in a good year. After turning yellow or russet in the fall, the leaves remain on the tree all winter long, thus providing good winter cover for birds and squirrels.

Shingle Oak could easily be one of the finest oaks for landscape plantings if it were more available from nurseries.

Quercus imbricaria · Shingle Oak

habitat . . .
FOREST. upland mesic-dry and lowland wet mesic, rocky uplands, floodplain and stream edge
- zone – 5a

form . . .
CONICAL. narrow, symmetrical crown, large understory tree (30–45′)
- branching – DESCENDING to horizontal limbs
- twig – smooth, GRAY-GREEN, slender with clustered, globular end buds; chestnut brown
- bark – scaly, GRAY-BROWN, slightly fissured into broad ridges

foliage . . .
ALTERNATE. simple, elliptic, shiny dark green LEATHERY leaves (4–6″) with entire, wavy margin; apex bristly, hairy below; whitish brown
- color (fall) yellow-BROWN to red
- season – deciduous, PERSISTING through early winter

flower . . .
CATKIN. slender, YELLOW-green (2–3″), singular drooping clusters; staminate
- sex – monoecious

fruit . . .
ACORN. tan, broad, DOME-shaped (½–¾″), ⅓–½ enclosed in bowl-shaped cup; thin, RED-BROWN scales
- season – maturing second year

OAK-HICKORY COMMUNITY

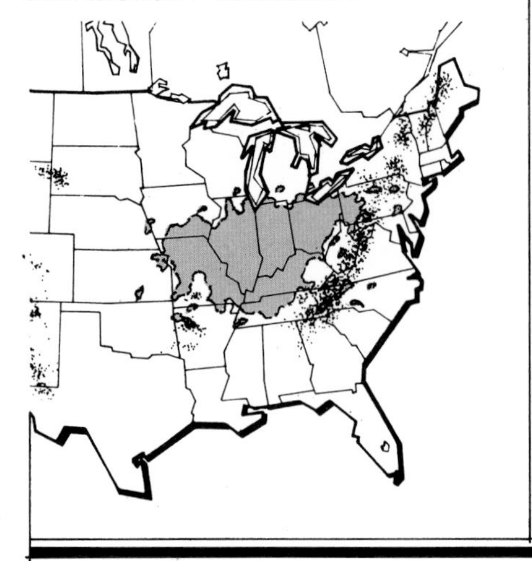

Chinkapin Oak, or Yellow Oak, is an irregularly shaped tree, generally as wide as it is tall. It is an upland tree that grows naturally on dry slopes and rocky limestone bluffs, but it attains its best growth in the rich soils in the lower Ohio Valley (Grimm, 1983).

The name comes from *chechinkamin,* the Algonquian Indian word for chestnut tree, meaning great berry (Borland, 1983). The leaves of the Chinkapin Oak resemble those of the chestnut, and the acorns are sweet and edible, as are chestnuts. Chinkapin Oak is sought by such wildlife as deer, squirrels, raccoons, and wild turkeys because it drops large crops of acorns every three or four years.

Chinkapin Oak grows 100 or more feet tall in some localities, but normally it reaches a height of less than 50 feet. It is a slow-growing tree that can live for 150 to 200 years. Like most oaks, it is difficult to transplant and therefore is seldom found in home gardens.

habitat . . .
SAVANNA. upland dry, steep rocky bluffs and dry stream banks
- zone—4b

form . . .
GLOBULAR. large understory tree (35–50′)
- branching—irregular, short, ASCENDING limbs
- twig—slender, GRAY-brown, glabrous with clustered, sharp-pointed end buds; orange-brown
- bark—ASH gray with thin, scaly, shallow-furrowed, flaky ridges

foliage . . .
ALTERNATE. simple, oblong-lanceolate, shiny bluish green leaves (4–7″) with COARSE-toothed margin; straight, parallel side veins, lustrous above, pale below; hairy
- color (fall) YELLOW-brown
- season—deciduous

flower . . .
CATKIN. YELLOW-green, singular drooping (2–3″) clusters; staminate
- sex—monoecious

fruit . . .
ACORN. glossy, tan-brown, ovoid DOME-shaped (½–¾″), ½ enclosed in scaly BOWL-shaped cup; scales long-pointed
- season—maturing first year

OAK-HICKORY COMMUNITY

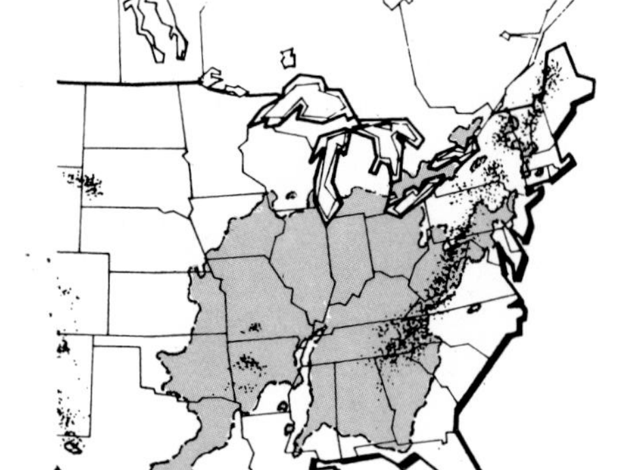

Quercus muhlenbergi · Chinkapin Oak

Black Oak is a tall, rugged tree that grows in the dry, gravelly, upland soils usually frequented by the White Oak Family. It lends its name to a large group of oaks with many similar characteristics. Other trees in the Black Oak Family are Red Oak, Pin Oak, Northern Pin Oak, Shingle Oak, and Scarlet Oak. These trees all have bristle-tipped leaves, bark that is dark in color, acorns that mature every other year, and buds that are pointed instead of blunt.

Black Oak, because of its name, must be the standard to which all other oaks are compared. Certainly it gives the appearance of having been around since the beginning of time. The old, outer bark is thick and deeply grooved. The leaves in autumn have no vivid colors, only dull red at best but mostly brown.

Black Oak normally is long-lived and relatively disease free. It grows sufficiently fast to stand up well among desirable shade trees for home plantings.

Quercus velutina · Black Oak

habitat . . .
FOREST, SAVANNA. upland mesic-dry and dry, sand, gravelly ridges and upper slopes
- zone – 4b

form . . .
OVOID to irregular. large canopy tree (75–100′)
- branching – symmetrical, ASCENDING limbs
- twig – stout, RED-brown with conspicuous lenticels having sharp-pointed, clustered buds; woolly scales
- bark – thick, nearly BLACK, flat ridges, shallow to deep-furrowed; orange inner bark

foliage . . .
ALTERNATE. simple, pinnately LOBED (7–9) glossy dark green leaves (5–8″) with bristly, lobed tips; lobes with deep sinuses, coppery below
- color (fall) yellow, GOLDEN brown
- season – deciduous, PERSISTENT through mid-winter

flower . . .
CATKIN. slender, YELLOW-green singular drooping (2–3″) cluster; staminate
- sex – monoecious

fruit . . .
ACORN. tan-brown, ELLIPTIC (¾″), ½–¾ enclosed by deep, HAIRY, red-brown cup; scales fringed
- season – maturing second year

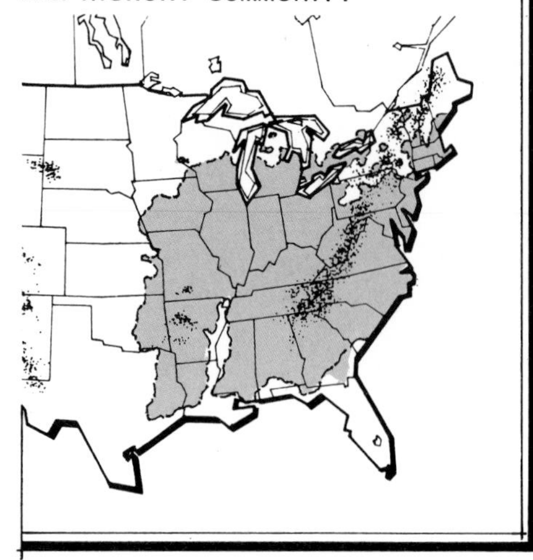

Ohio Buckeye is a medium-sized canopy tree, 50 to 70 feet tall. It is usually found mixed among members of the Oak Family on dry, south- and west-facing slopes, but never in great abundance. It is commonly used as an ornamental shade tree because of its interesting fruit and bright orange fall color.

Few trees are so aptly named as is Buckeye, for its large brown nuts do suggest the eyes of a deer. It is a very distinctive tree with thick-husked nuts, large palmately compound leaflets grouped in clusters like the fingers of a hand, and prominent, often showy, erect blossom clusters held at the ends of the twigs.

Ohio Buckeye would be an excellent street tree with one possible limitation: its litter. We enjoy a tree that is constantly contributing to change on the ground surface, but there are some who find Ohio Buckeye too messy.

habitat . . .
FOREST. upland mesic-dry, lowland wet-mesic, floodplain terrace, south and west facing slopes
- zone—4a

form . . .
IRREGULAR. broad, large understory tree (35–50′)
- branching—limbs stout with RECURVING branches
- twig—smooth, BROWN to black, stout with large papery (5/8–1″) buds, EXFOLIATING red-brown; terminate
- bark—BROWN-black, deeply fissured with thin, irregular scales

foliage . . .
OPPOSITE. palmately COMPOUND, leaflets (5) long-stemmed, lanceolate yellow-green leaves (2–6″) with finely serrated margin; pale below
- color (fall) ORANGE to red
- season—deciduous

flower . . .
SPIKE. small, yellow (1/2″), TRUMPET-shaped, (2–5″) erect pyramidal clusters; terminal
- sex—monoecious, appearing after leaves

fruit . . .
CAPSULE. smooth, GLOSSY red-brown nut; encased by spiny (1–2″) husk; splitting
- season—maturing summer, early autumn

OAK-HICKORY COMMUNITY

Aesculus glabra · Ohio Buckeye

Flowering Dogwood is found in rich, moist, well-drained soils in countless woodlands throughout the eastern half of the United States.

At every season of the year, Flowering Dogwood is an exceptional tree. It is best known and appreciated in spring, for then the four large, white to pink-tinged bracts of its flower unfold, as the Dogwood flowers fill the woods with drifts of white. In summer, the tree is handsomely clothed with two-toned leaves, green above and silver below. In fall, the rich wine tone of the leaves provides a striking background for the bright red fruit clusters that are enjoyed by more than one hundred species of birds. The tree's delicate horizontal branching becomes apparent with its winter silhouette, and its gray button buds are promises of blossoms in the spring.

Flowering Dogwood makes an excellent and useful tree for home landscaping, where it does best in a protected location. It grows slowly, lives a long time, and does well within its native range.

habitat . . .
FOREST. upland mesic, mesic-dry, ravines, bluffs and wooded slopes
- zone – 5b

form . . .
GLOBULAR. dense, often FLAT-topped, large understory tree (35–50′)
- branching – horizontal, often STRATIFIED with branchlets upturned at twig ends
- twig – smooth, green to purplish red, stout twigs encircled by conspicuous leaf scar with stalked, ONION-shaped terminal buds; FLESH color
- bark – gray, BLOCKY with square, grid-like segments

foliage . . .
OPPOSITE. simple, elliptic-oval, bright yellow-green leaves (3–6″) with somewhat WAVY entire margin; parallel (5–8) veination along margin, apex taper-pointed, thick, whitish beneath
- color (fall) scarlet, MAROON-red
- season – deciduous

flower . . .
CLUSTER. small, yellow, surrounded by (4) WHITE bracts (1–1½″) across, compact clusters appearing as single flower; bracts parallel-veined, blunt to NOTCHED apex
- sex – monoecious, FRAGRANT

fruit . . .
BERRY. small, scarlet-RED (⅜″), FOOTBALL-shaped, glossy clusters
- season – maturing late summer, early autumn

(marginal throughout Iowa)
OAK-HICKORY COMMUNITY

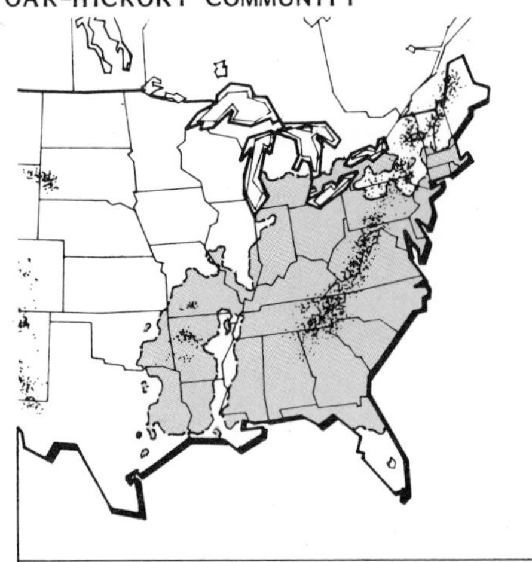

Cornus florida · Flowering Dogwood

American Hophornbeam, or Ironwood, is a small tree that is very adaptable and independent. Common throughout the eastern half of the United States, it grows in many different soils. It is tolerant of shade and can be found almost anywhere in the woods. Hophornbeam always grows by itself, never in groves or other groupings, so many people completely overlook it.

Because of its leaf size and shape, American Hophornbeam is often assumed to be an elm. Upon closer examination, however, Hophornbeam can be easily recognized. First, the bark is very distinctive. On a mature tree, it is thin and flaky and broken into small flat scales, which are generally loose at both ends and fastened tightly in the middle. There is also a slight twist in the bark on the trunk, giving the appearance of a large hemp rope.

Then there are the catkins, usually appearing in groups of three at the ends of the delicate, finely textured twigs. Finally, the fruit, which matures in the fall, resembles hops in its small clusters of overlapping, flattened sacs, each containing a flat nut. These nuts are eaten by many birds, including ruffed grouse, and they are enjoyed by squirrels as well.

The wood, as the name implies, is as hard as horn and as strong as iron. It is used commercially for tool handles, mallets, and fence posts. American Hophornbeam is a slow-growing tree that is long-lived and is rather difficult to transplant.

habitat . . .
FOREST. upland mesic and upland mesic-dry, upland slopes, ravines and rocky streams
- zone – 5a

form . . .
CONICAL. open, large, symmetrical understory tree (35–50′)
- branching – IRREGULAR to horizontal limbs, slightly pendent
- twig – smooth, glossy RED-brown, slender ZIG-ZAG with bud at angle having 3 bundle scars
- bark – scaly, gray-BROWN, loose, narrow strips of SHREDDY appearance

foliage . . .
ALTERNATE. simple, oval to elliptic, yellow-green leaves (3–5″) with DOUBLE-toothed margin; parallel veins with uneven base, somewhat hairy below
- color (fall) YELLOW-brown
- season – deciduous, PERSISTING until February

flower . . .
CATKIN. (male) red-BROWN, slender, (½″) cylindrical in GROUPS of 3 at twig end; (female) REDDISH green, small (½–¾″) cluster in leaf axils
- sex – monoecious

fruit . . .
SAMARA. tan-brown, SAC-like (1–1½″), flattened nutlet, suspended cluster of HOPS; pendulous
- season – maturing late summer

Ostrya virginiana · American Hophornbeam

Common Buckthorn is a tall shrub introduced from Europe that has become widely naturalized by birds scattering its seeds. The shrub tolerates a wide variety of soils and grows equally well in sun or shade.

The thorn of Common Buckthorn is neither obvious nor of much protective value to the plant, for it grows only at the tips of the branches, emerging between two dark brown buds. The flowers are also relatively inconspicuous but very fragrant, blooming yellow-green as the plant leafs out.

Male and female plants are separate. The fruit on the female shrub is the most striking feature of Common Buckthorn. The shrub produces a profuse crop of black berries that mature in September. Common Buckthorn is an attractive ornamental for its dark green, shiny leaves and clusters of black fruit.

It is an excellent shrub for hedges and background plantings. While it holds its leaves longer in the fall than most other plants, the leaves remain green and have no special fall color. Several improved buckthorns, such as Tallhedge, are available from nurseries.

habitat . . .
FOREST. upland mesic-dry and upland dry, dry soils, clearings and open woods
- zone—3a

form . . .
OBOVOID. erect, small understory tree (12–20′)
- branching—UPRIGHT, bushy topped with spiny branches
- twig—slender, dark GRAY, paired, unlined ending in SHARP spines; buds narrow, scaly pointed, often SUB-opposite
- bark—BROWN-black, smooth, curly strips; somewhat warty

foliage . . .
OPPOSITE. simple, elliptic to ovate, dull green leaves (1–2½″) with fine-toothed margin; base broadly wedge-shaped on SHORT spurs; 3-veined
- color (fall) GREEN
- season—deciduous, PERSISTING through fall

flower . . .
CLUSTER. small, yellow-green, PERFECT (⅛–¼″), bell-shaped, inconspicuous clusters in leaf axils
- sex—dioecious

fruit . . .
BERRY. small, BLACK, (¼″) dia. in pairs or small cluster; often persisting
- season—maturing late summer and autumn

(Introduced from Europe)
OAK-HICKORY COMMUNITY

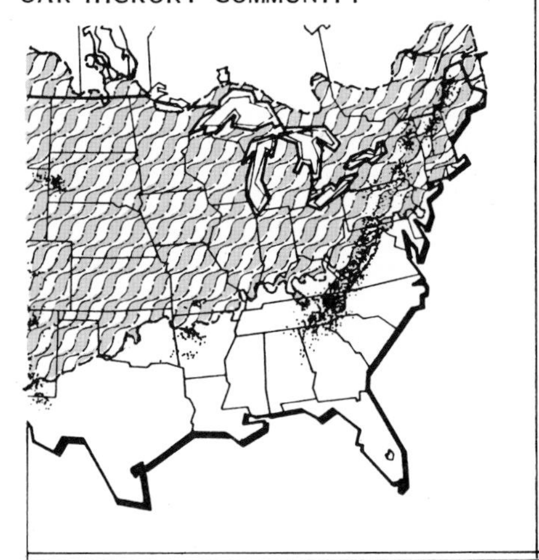

Rhamnus cathartica · Common Buckthorn

Nannyberry Viburnum is one of the tallest viburnums, occasionally growing as a small tree, 20 to 25 feet tall. It tolerates both sun and shade and grows well in any good soil.

It is easily identified in winter by its long, tapered, terminal flower buds, which are swollen at the base. In the summer, Nannyberry Viburnum can be recognized by its leaf stems, which have a winged margin. In fall, the edible blue-black berries are helpful in identification, for they have a thin, sugary flesh covering a large seed stone. (Blackhaw Viburnum produces similar fruit.) Like other viburnums, Nannyberry Viburnum has attractive white flowers in spring, dense foliage in summer, and excellent wine-red leaves in the fall.

Viburnums in general are highly recommended for a natural setting or shrub border, for they have something special to offer throughout every season of the year.

habitat . . .
FOREST, SAVANNA. upland mesic-dry and dry, stream edge, wooded slopes and woods edge
- zone – 2

form . . .
OBOVOID. dense, small understory tree (20–35′)
- branching – UPRIGHT, spreading limbs
- twig – slender, gray-brown with naked FLESH color, long-pointed, terminal bud; swollen base, hairy
- bark – blocky, BROWN-black, giving a squarish, scaly, PLATE pattern

foliage . . .
OPPOSITE. simple, ovate-oblong ovate, glossy bright green leaves (1½–2″) with fine-toothed margin; base WEDGE-shaped with winged petiole, pointed apex having prominent veination
- color (fall) PURPLISH, red-orange
- season – deciduous

flower . . .
FLAT-TOPPED. creamy WHITE (3–4″) across terminal clusters; FRAGRANT
- sex – monoecious

fruit . . .
BERRY. open, RED, bluish black, (⅜″) dia., loose, drooping clusters on bright RED stems
- season – maturing late summer, PERSISTING through winter

OAK-HICKORY COMMUNITY

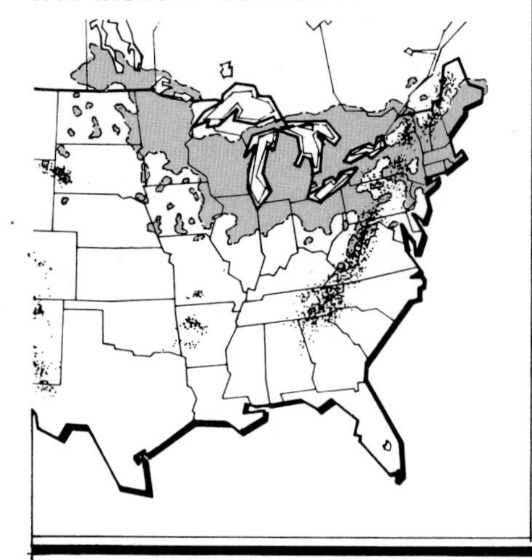

Viburnum lentago · Nannyberry Viburnum

Tatarian Honeysuckle was introduced to this country from Europe. It has become so well naturalized throughout the woods and countryside that it now grows like a native shrub. It thrives on a variety of soils and adapts to sun or shade.

Tatarian Honeysuckle is a popular plant for hedges and windbreaks because it grows fast and produces attractive flowers and fruit. The flowers are white, pink, or red and are composed of five petals that unite in a funnel or trumpet-shaped tube. They are fragrant and are pollinated by insects and occasionally hummingbirds.

While the flowers are attractive, they do not compare with the colorful fruit—pairs of red berries fused together. The fruit is produced in great abundance in late June and July. Although eaten by many birds, the fruit does not seem to become less abundant, and it adorns the bushes for a considerable time. Children are often tempted to eat this berry but are usually discouraged by the flavor, which is bitter and lingers in the mouth.

Tatarian Honeysuckle tends to grow into a leggy form as a mature plant. The leaves, as on all honeysuckles, are opposite. They maintain a deep green color late in the season and lack a special change of color in the fall. Other varieties are now available from nurseries that maintain a fuller shape.

habitat . . .
FOREST, SAVANNA. upland mesic dry, open woods, open fields and roadsides
- zone—3b

form . . .
GLOBULAR. dense, large shrub (8–10′)
- branching—upright, TWIGGY with hollow branchlets
- twig—papery, gray, hairless; brown PITH with short, blunt buds
- bark—tan-brown, PAPERY; bushy

foliage . . .
OPPOSITE. simple, ovate-oblong ovate, green leaves (1–2½″) with entire margin; base slightly HEART-shaped, pale below, hairless
- color (fall) YELLOW-green
- season—deciduous

flower . . .
BELL-shaped. pink (¾″), 2-LIPPED on long, slender stalk; paired
- sex—monoecious

fruit . . .
BERRY. small, RED (⅛″), globular, slightly UNITED at base
- season—early to mid-summer

(Introduced from Europe)
OAK-HICKORY COMMUNITY

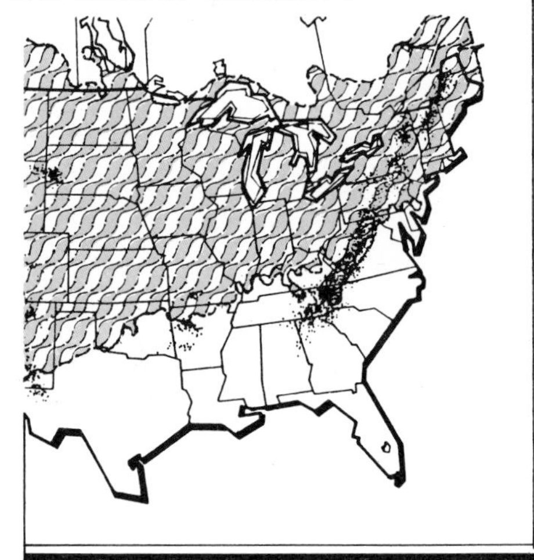

Lonicera tatarica · Tatarian Honeysuckle

Common Pricklyash is a many-branched, upright shrub that often forms dense thickets and is common to dry soils in oak woods or at the edges of woods.

Common Pricklyash is a citrus, and all parts of the shrub are pungent and aromatic. When the leaves are crushed, they give off a strong lemony odor. The larvae of the Giant Swallowtail Butterfly feed only on citrus; therefore, Common Pricklyash allows this attractive insect to expand the northern part of its range.

The rather inconspicuous flowers appear early and are greenish-white. The fruit ripens in the fall into a small ⅛-inch capsule that contains a solitary, lustrous, black seed. Both the fruit and bark have been used for home remedies. If chewed, they are supposed to cure a toothache, but the taste left in one's mouth is almost as bad as the toothache itself.

The name, Pricklyash, comes from the stems, which are armed with nodal pairs of broad-based thorns. The leaves are compound, somewhat resembling an ash. However, they are arranged alternately along the stem rather than being opposite, as the ashes are. The yellow autumn color of the leaves is very attractive.

Common Pricklyash, with its many interesting and novel characteristics, can add excitement to landscapes and urban planting designs.

habitat . . .
FOREST, SAVANNA. upland mesic-dry and dry, open rocky slopes, open fields and woods edge
- zone – 3a

form . . .
OBOVOID. irregular, small understory tree (12–25′)
- branching – dense, ERECT, ascending limbs
- twig – smooth, gray-brown, stout, armed with short, paired spines flanking globular, red buds; woolly
- bark – smooth, gray-reddish brown becoming broken, fine scales (old)

foliage . . .
ALTERNATE. pinnately COMPOUND, leaflets (5–11) ovate-elliptic, green LEATHERY leaves (1–3″) with dull, toothed margin; leaflets sometimes prickly
- color (fall) greenish YELLOW
- season – deciduous, LEMON-like odor when crushed

flower . . .
CLUSTER. small, yellow-GREEN (⅛–¼″), with spreading petals; short-stalked
- sex – dioecious, appearing before leaves

fruit . . .
CAPSULE. fleshy, RED turning to black, (⅛″) splitting pod in small dense clusters; seeds black
- season – maturing late summer, early autumn

OAK-HICKORY COMMUNITY

Zanthoxylum americanum · Common Pricklyash

Fragrant Sumac is a low-spreading shrub that normally grows 4 to 6 feet tall but may reach 8 feet in height. It is common to dry, well-drained soils of oak woods, clearings, and the edges of woods. It is called Fragrant Sumac because of the leaves, which give off a pleasant, aromatic odor when crushed.

Fragrant Sumac is a popular ornamental shrub for massing effects and specimen plantings because of its attractiveness throughout the seasons. In winter, Fragrant Sumac is a rather sparse, gray bush containing many small flower buds that look like catkins. These buds open in May, loading the branches with yellow flowers. The dark green foliage of summer is handsome, but in no way compares to the autumn splendor that follows. Fragrant Sumac's leaves turn brilliant hues of red, yellow, and orange, all on the same shrub. The fruit, too, is attractive. Clusters of dense, hairy red drupes sometimes remain on the plant all year, furnishing food for birds.

All in all, Fragrant Sumac is an excellent plant for urban and rural landscapes. It is put to best use when planted in mass or drift-like plantings as it occurs in nature.

Rhus aromatica · Fragrant Sumac

habitat . . .
SAVANNA. upland mesic-dry, open upland woods, cliffs and rocky bluffs
- zone – 3

form . . .
GLOBULAR. irregular, multiple-stemmed, erect shrub (6–12′)
- branching – spreading, HORIZONTAL with dense, ascending outer limbs
- twig – smooth, BROWN, with leaf buds hidden by ROUND leaf scar; pubescent, false end bud
- bark – rough, brown, shaggy appearance

foliage . . .
ALTERNATE. trifoliately COMPOUND, leaflets (3) ovate, dark green leaves (1½–3″) with coarse-toothed margin; narrow base on short leaf stalk, shiny above, FUZZY below
- color (fall) RED, orange-yellow
- season – deciduous, AROMATIC

flower . . .
CATKIN. small; (female) yellow (¼–½″), dense clustered spikes in GROUPS of 3; (male) brown
- sex – dioecious

fruit . . .
BERRY. small, RED, globular, (¼″) dia., terminal massed clusters; VELVETY
- season – maturing late summer

OAK-HICKORY COMMUNITY

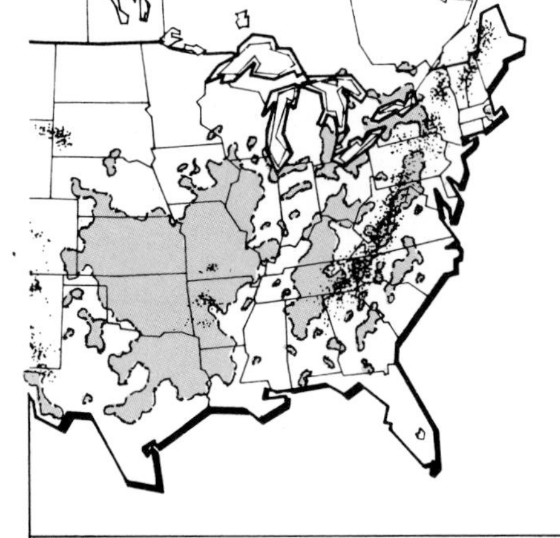

Pasture Gooseberry is a small thorny shrub that tolerates shade and poor soils. It is the most common, most widely distributed, and best known of the wild gooseberries. The greenish-white, bell-like flowers bloom from late April to June. The fruit ripens during July and August. In spite of the fruit's prickly appearance, it has a good flavor when cooked, and it is often used for making preserves, jellies, and pies.

Pasture Gooseberry is also called Prickly Gooseberry for good reason. Not only are the berries prickly, but the plant defends itself with many prickles on the lower part of its stems. The upper stems are entirely smooth except for the spines at the base of the leaves. This thorny plant is a frequent nesting site for several songbirds, and its fruit is relished by birds of many species.

Pasture Gooseberry is useful in shrub borders, but it is an alternate host for five-needle pine blister rust and should not be planted near White or Limber Pine.

habitat . . .
FOREST. upland mesic-dry, dry, wet-mesic, stream and creek banks, rocky slopes and old clearings
- zone–2

form . . .
GLOBULAR. upright, dense small shrub (3–5′)
- branching–spreading, ARCHING stems, often straggling appearance
- twig–slender, brown, armed with 3-parted red (1/4–3/8″) SPINES with surrounding small reddish brown bud
- bark–tan-brown, EXFOLIATING to shallow-furrowed with dense, PRICKLY spines

foliage . . .
ALTERNATE. simple, palmately (3) LOBED, bright green leaves (1–2″) with irregularly toothed margin; soft and hairy below
- color (fall) yellow, orange-REDDISH purple
- season–deciduous

flower . . .
BELL. slender, greenish white (3/8–1/2″), TRUMPET-shaped clusters of 1 to 3; PENDENT
- sex–dioecious

fruit . . .
BERRY. globular, red-PURPLE, (1/2″) dia. covered with tiny spines; PRICKLY
- season–maturing late summer

OAK-HICKORY COMMUNITY

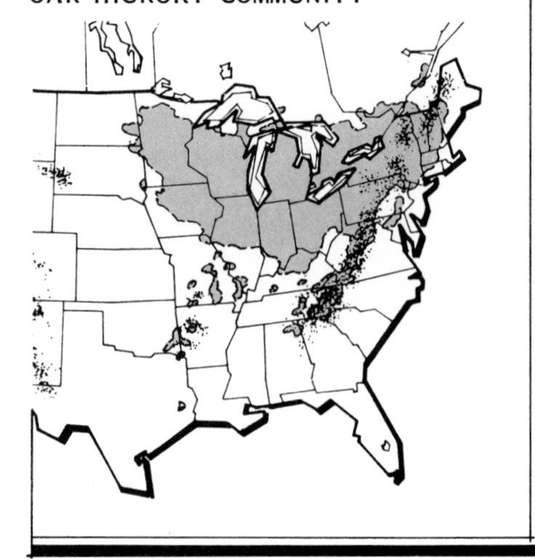

Ribes cynosbati · Pasture Gooseberry

Rafinesque Viburnum, or Shortstalk Viburnum, is an excellent plant for a shrub border. Like most viburnums, it has white flowers in spring, dense foliage in summer, and attractive coloration in the fall.

Rafinesque Viburnum is a small, bushy shrub, 2 to 5 feet tall. It prefers to grow in dry areas and is often found on stony slopes and banks. It tolerates sun or partial shade and grows in any good garden soil.

Its name, Shortstalk, comes from the plant's short petioles. The upper surface of the leaf is smooth, but the lower surface is soft and downy, giving it another name, Downy Arrowwood. The leaves of Rafinesque Viburnum have prominent leaf margins with coarse teeth.

Like many other viburnums, Rafinesque Viburnum has flat to round-headed, showy, white flowers followed by clusters of small, shiny, blue-black berries that provide valuable food for birds and other wildlife. Also, its foliage has beautiful fall color, ranging from yellow-orange to deep wine-red.

Rafinesque Viburnum can be used on any site with dry slopes where it is difficult to establish plant material. It is an excellent plant for a shrub border or massing effect because of its attractiveness throughout the seasons.

habitat . . .
FOREST. upland mesic-dry, rocky slopes, bluffs and wooded clearings
- zone—3a

form . . .
GLOBULAR. narrow, open small shrub (3–6′)
- branching—slender, MULTIPLE-stemmed, spreading branches often in loose habit
- twig—slender, GRAY-brown, smooth with small (1/8″) rounded buds; red-brown
- bark—smooth, gray-brown with conspicuous LENTICELS (old) shallow-furrowed

foliage . . .
OPPOSITE. simple, ovate, dull green leaves (1½–3″) with coarse, DENTATE-toothed margin; pronounced parallel veination
- color (fall) deep MAROON-purple
- season—deciduous

flower . . .
FLAT-TOPPED. small, creamy WHITE (1/4″) clusters (2–3″) across; terminal
- sex—monoecious, appearing after leaves

fruit . . .
BERRY. smooth, bluish black, (1/4–3/8″) dia., OVAL-shaped in flat clusters
- season—maturing late summer, early autumn

OAK-HICKORY COMMUNITY

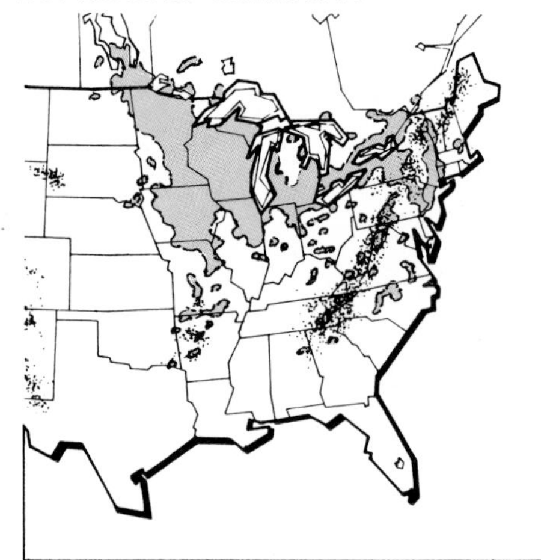

Viburnum rafinesquianum · Rafinesque Viburnum

American Bittersweet is a vigorous vine that scrambles over the ground or low vegetation and often climbs by twining 20 or more feet through the branches of trees. It grows in sun or partial shade in a variety of soils, but it prefers a drier site such as oak woods or fencerows.

American Bittersweet has inconspicuous greenish-white flowers that appear in June. In the fall as its leaves turn yellow, Bittersweet produces very attractive fruit clusters. The yellowish husk splits to expose a bright red berry that keeps for a very long time.

The clusters of fruit and twigs are often gathered and sold for decorative uses. There are few native shrubs or vines as well known as American Bittersweet.

habitat . . .
FORESTᐧ mesic, mesic-dry, dry, rocky slopes, roadsides, woods and stream edge
- zone – 2

form . . .
CLIMBING. sometimes groundcover-like (3–18′)
- branching – TWINING stems often on itself, picturesque
- twig – slender, SMOOTH, yellow, gray-brown having prominent LENTICELS with small, conical knob buds; red-brown
- bark – smooth, gray-brown (old) trunks with FLAKE-like scales

foliage . . .
ALTERNATE. simple, elliptic to ovate, yellow-green leaves (1½–3″) with fine-toothed, wavy margin; glossy above, paler beneath
- color (fall) lemon-YELLOW
- season – deciduous

flower . . .
CLUSTER. small, yellow-green, (⅛–⅜″) dia., terminal clusters; inconspicuous
- sex – dioecious, appearing after leaves

fruit . . .
CAPSULES. bright orange, 3-SECTIONED, (⅛–½″) dia., loose, drooping clusters with SCARLET-red seeds
- season – maturing in autumn

OAK-HICKORY COMMUNITY

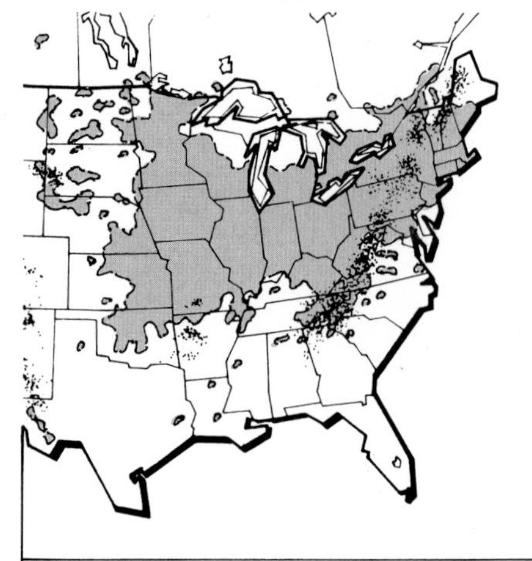

Celastrus scandens · American Bittersweet

Limber Honeysuckle is a low, climbing, vine-like shrub with branches arching or twining 3 to 10 feet from the plant. It is found on sloping land in dry oak woods, and it grows in either shaded woods or sunny clearings.

Limber Honeysuckle has distinctive red tubular flowers with yellow stamens, and it blooms early in May. The flowers are arranged in whorls at the end of the arching branches. Later in the summer, its red berries ripen in terminal clusters. Like most honeysuckle berries, they are bitter but not toxic. Birds, however, relish this fruit and widely disseminate the seeds.

The most unusual characteristic of Limber Honeysuckle is its leaves. Like all honeysuckles, its leaves are oppositely arranged on the stem, but the uppermost pair of leaves on each stem is perfoliate (joined around the stem as a disk), so the stem appears to pierce the leaf. This leaf arrangement provides a collar background for both bloom and fruit.

Limber Honeysuckle is an exceptional little shrub that provides much interest in natural gardens, especially when planted with other early-blooming forest wildflowers.

habitat . . .
FOREST. mesic-dry, dry, rocky woods, bluffs and sandstone slopes
- zone–2

form . . .
UPRIGHT. open, single-stem shrub (1–3′)
- branching–singular, ARCHING to straggling vine
- twig–smooth, tan, GRAY-brown with small, pointed (1/4″) buds; tan-brown scales
- bark–thin, gray-brown with EXFOLIATING, vertical stripes, fibrous

foliage . . .
OPPOSITE. simple, narrow, elliptic or oval, medium green leaves (1–2 1/2″) with entire margin; upper pairs unite at base ENCIRCLING stem; pale, downy, white beneath
- color (fall) YELLOW-green
- season–deciduous

flower . . .
TUBULAR. 2-lobed, purplish yellow-orange (1/2–1″) with bright yellow stamens in dense whorls
- sex–monoecious, FRAGRANT

fruit . . .
BERRY. small, red, globular, (1/8–1/4″) dia., terminal clusters; whorled
- season–mid to late summer

OAK-HICKORY COMMUNITY

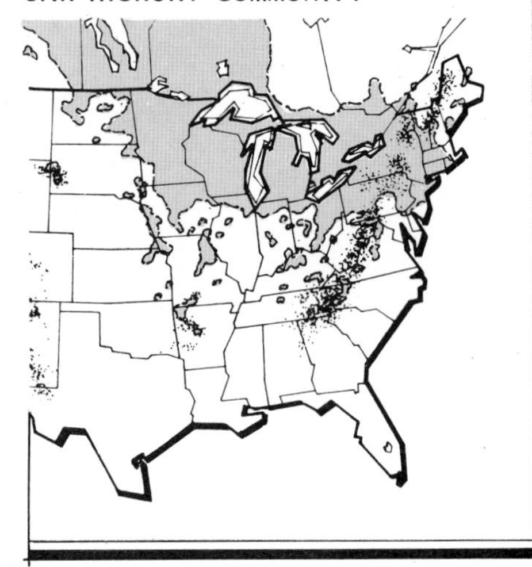

Lonicera dioica · Limber Honeysuckle

The Maple-Linden community is the western equivalent of the Maple-Beech community that extends eastward to the Allegheny Mountains. The natural range of American Beech does not extend to the northwestern limits of the deciduous forest. Beech is replaced by American Linden, beginning in Wisconsin and continuing into Minnesota and Iowa (Oosting, 1956). The Maple-Linden community of the Midwest is only found on sloping land that faces north and east.

Here, hot, dry periods restrict this community to small pockets and protected coves that receive and retain sufficient soil moisture. Many trees would grow in this desirable microclimate of high humidity and cool, stable air at surface levels. However, there are few that can germinate in the dense shade of the Maple-Linden canopy. Occasionally an oak is found competing for sunlight, but unless it is a Red Oak that can germinate in these low-light conditions, the development of future generations is unlikely.

The moist ground layer is especially favorable to numerous wild flowers and ferns. The low light level, however, causes plants to emerge, bloom, seed, and wither in only a few short weeks before the leaf cover of the dense Maple-Linden canopy blots out the sun.

Maple-Linden Community

Trees: Canopy (over 48′)

Trees: Understory (12–48′)

Shrubs: (6–12′)

Shrubs: (less than 6′)

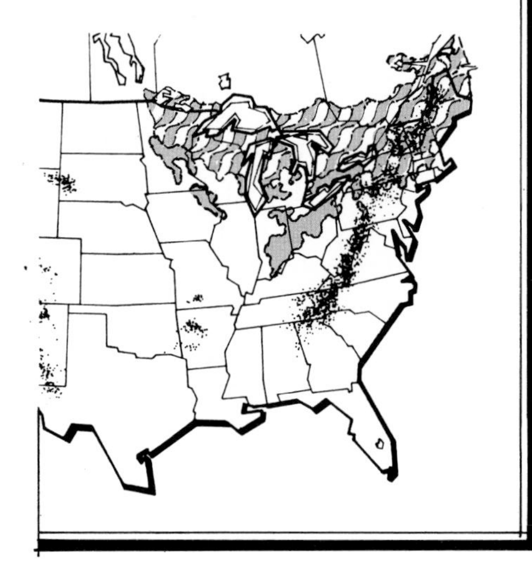

Black Maple prefers the deep, moist soils of upland forests. It is a stately tree with a rounded crown very similar to Sugar Maple's. It replaces Sugar Maple in most of Iowa and is the only hard maple in South Dakota (Sargent, 1961).

Black Maple is easily recognized in summer with its three-lobed, thick, drooping leaves. The leaves are a darker green than are Sugar Maple's, and the leaf margins are not indented but have a few wavy teeth. Often there are prominent stipules at the base of the leaf stalks. In all seasons, Black Maple may be distinguished by the bright orange color of its newest twigs.

For those interested in maple sugar production, it has been said by specialists that a Black Maple tree yields more and sweeter sap than does a Sugar Maple of the same age (Werthner, 1935). Black Maple's wood is much like that of Sugar Maple; in fact, both are marketed as hard maple.

There are other minor differences in the bark and fruit, but for the most part the excellent features that Black and Sugar Maple have in common make them equally desirable for shade trees in lawns or street plantings. They are both moderate- to slow-growing trees that live 200 to 300 years.

Acer nigrum · Black Maple

habitat . . .
FOREST. upland mesic, coves, mesic ravines, north and east facing slopes
- zone – 3b

form . . .
OVOID. globular, large canopy tree (75–100′)
- branching – wide SYMMETRICALLY ascending limbs; dense
- twig – GLOSSY, yellow-brown, slender, marked by conspicuous LENTICELS with clustered, sharp-pointed end buds; black, HIDDEN by the leaf stem
- bark – brown nearly black, deep-furrowed with long SCALY plates

foliage . . .
OPPOSITE. simple, palmately 3-LOBED, yellow-green leaves (4½–5″) with often DROOPING edges; dull, lobes long-pointed; tomentose beneath and petiole
- color (fall) yellow-ORANGE, RED
- season – deciduous

flower . . .
CLUSTERS. small, yellow-green, (3/8″) dia., BELL-shaped with slender, STRINGY filaments; pendulous
- sex – dioecious

fruit . . .
SAMARA. tan-brown, PAIRED (1–1¼″) key in loose clusters, one seeded; pendulous
- season – maturing late summer, early autumn

MAPLE-LINDEN COMMUNITY

Sugar Maple is unsurpassed for splendid autumn foliage, for it has a spectacular blend of reds, golds, and greens. As a shade tree, it is far better than either Norway Maple, which is too dense, or Silver Maple, which is too brittle.

Sugar Maple is found on cool, moist slopes in eastern forests, but it grows just as well in any productive, well-drained location.

Sugar Maple can be moved from the woods when it is a small tree, but nursery-grown trees are best for ease of transplanting and symmetry of shape. Care should be taken in either case to leave the lower limbs and to wrap the trunk each fall until the tree is an established, mature specimen. This will prevent sunscalding of the bark.

Sugar Maple's pale, close-grained wood is highly valued for cabinetmaking. Abnormal growth patterns make the wood equally popular for use as bird's-eye maple interior paneling. Sugar Maple is also well known for its supply of sweet sap, which is used to produce maple sugar.

habitat . . .
FOREST. upland mesic, mesic ravines, north and east facing slopes
- zone – 3a

form . . .
OVOID. globular, large canopy tree (75–100′)
- branching – dense, ASCENDING limbs
- twig – slender, gray-brown with slender, long-pointed, clustered buds; gray-brown, NOT hidden by the leaf stem
- bark – gray-brown to black, deep-furrowed with scaly ridges, major LIMBS black

foliage . . .
OPPOSITE. simple, palmately 5-LOBED, bright green leaves (3½–6″) with narrow, long-pointed tooth margin; long petiole with moderate sinuses
- color (fall) ORANGE-red to yellow
- season – deciduous

flower . . .
CLUSTER. small, yellow-green, (⅜″) dia., BELL-shaped with slender STRINGY filaments; pendulous
- sex – dioecious

fruit . . .
SAMARA. tan-brown, paired, U-shaped, (1–1¼″) KEY in loose cluster, one seeded; pendulous
- season – maturing in September

MAPLE-LINDEN COMMUNITY

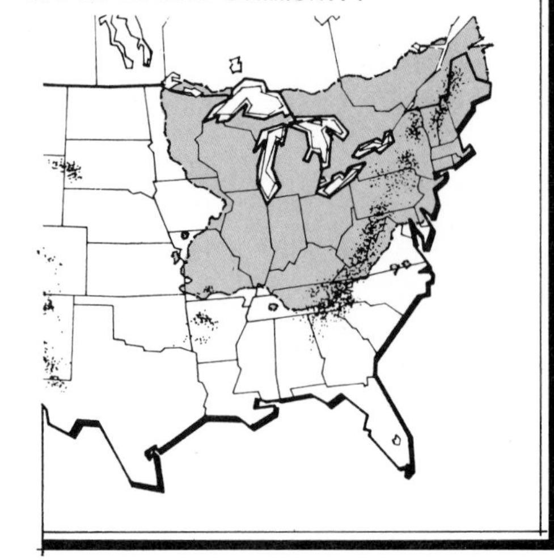

Acer saccharum · Sugar Maple

American Chestnut was once a tall, straight tree, occasionally reaching 100 feet in height with a trunk up to 8 feet in diameter. Now it exists primarily as sprouts from old stumps. American Chestnut was once common throughout Eastern Deciduous Forests and in parts of the Midwest. It grew everywhere from well-drained soils in valleys to rocky ridges. This was true until the early 1900s, when a Chestnut Blight that started in New York City became an epidemic. In less than a generation, American Chestnut was eliminated as an important forest tree.

Children of today have lost an important legacy: the fine sport of chestnut hunting on a crisp, autumn morning. After a frost, the fierce prickles of the husk would split, allowing the chestnuts to fall to the ground below. There was a great competition, searching among grass and fallen leaves for the biggest nuts and the greatest quantity of them. There was always a good nut crop, for American Chestnut did not produce blossoms until June or early July after all danger of frost was past, unlike oaks and beeches whose blossoms can be killed by late frosts.

There are still a few old Chestnuts left, like the one pictured, that somehow became isolated and were not killed by the blight. But they are rare. So we must be content with photographs and memories of old trees and with saplings that grow from stumps until they reach 15 to 20 feet tall. Then the blight cuts them down. They are the "Peter Pans" of the tree world, for they never grow up.

Castanea dentata · *American Chestnut*

habitat . . .
FOREST. upland mesic-dry, upland mesic, dry ridges and rocky slopes
- zone—5a

form . . .
GLOBULAR. irregular, small canopy tree (50–75′)
- branching—massive, HORIZONTAL limbs
- twig—stout, RED-brown, glossy with 2-scale, clustered, long-pointed, lateral buds; chestnut brown
- bark—gray-brown, shallow, furrowed with broad FLAT ridges; ash-GRAY plates

foliage . . .
ALTERNATE. simple, oblong-lanceolate, dark yellow-green leaves (5–9″) with coarse-toothed, BRISTLY tipped margin; straight, parallel side veins; thin, smooth surface
- color (fall) yellow-BROWN
- season—deciduous

flower . . .
CATKIN. slender, yellow (6–8″), PENCIL-like spikes at base of leaf; (female) (3/8″) with greenish scales
- sex—dioecious

fruit . . .
NUT. reddish brown, (1/2–3/4″) dia. groups of 2s and 3s, incased by dense, SPINEY, bur-like, (2″) dia. husk; densely woolly inside
- season—maturing late summer

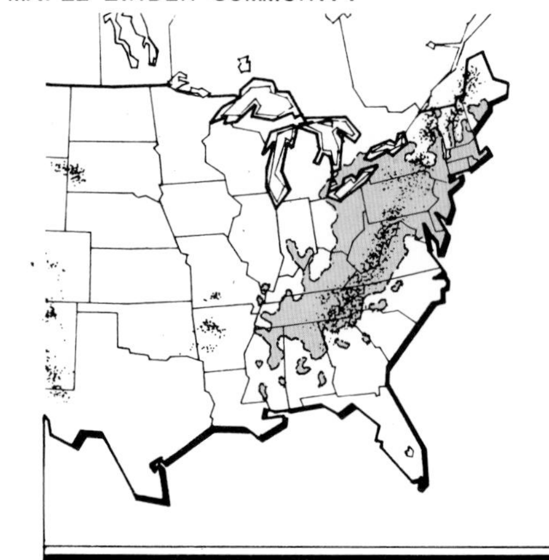

American Yellowwood is a small canopy tree that rarely reaches 50 feet in height. It has the distinction of being the Midwest's rarest native tree and the most local in its distribution. It was discovered on limestone cliffs that overhang streams in the moist, rich soils of Kentucky and Tennessee. Since then, it has been found locally in several other states, having been planted far beyond its natural range as a shade or ornamental tree.

American Yellowwood is generally as wide as it is tall, with a graceful crown of slender, pendulous branches that have gray bark as smooth as a beech's. American Yellowwood has two strange leaf characteristics. First, the leaves are compound and hold the leaflets in an alternate fashion. (Most trees with compound leaves hold their leaflets opposite each other along the central stalk.) Second, the base of the leaf stalk, or petiole, completely covers the bud, a condition common to only one other tree, the sycamore.

American Yellowwood is best known for its bloom, even though it seldom blooms in two consecutive years and rarely covers itself with flowers more often than two or three times a decade. Yellowwood is often planted as an ornamental tree, but it has weak crotches that may lead to splitting and breakage. It tends to grow at a medium-fast rate and may live to 120 years.

Cladrastis lutea· American Yellowwood

habitat . . .
FOREST. upland mesic, lowland wet-mesic, coves, ravines, limestone ridges and river valleys
- zone – 3b

form . . .
GLOBULAR. short trunk, small canopy tree (50–75′)
- branching – UPRIGHT, spreading limbs
- twig – glossy red-brown, smooth with prominent lenticels; leaf SCAR surrounding DOME-shaped buds; brown
- bark – thin, smooth, brown to ash gray

foliage . . .
ALTERNATE. pinnately COMPOUND, leaflets (7–11) oblong obovate, blue-green leaves (8–12″) with entire margins; pale beneath with leaf stalk enclosing bud
- color (fall) YELLOW
- season – deciduous

flower . . .
SPIKE. creamy white (12–14″), pyramidal, TERMINAL spike with PEA-like (1–1¼″) blooms; pendant
- sex – monoecious, fragrant

fruit . . .
LEGUME. tan-brown, FLATTENED-oblong (3–4″) pod in open clusters; pendulous
- season – maturing early fall, PERSISTING

MAPLE-LINDEN COMMUNITY

American Beech is a beautiful specimen tree that is often found in large groves. It is hardy, slow-growing, and long-lived. The bark is unique in that it is very smooth, light gray, and does not change as the tree grows older. The leaves are large and attractive with prominent veins. Fall coloration varies from orange to yellow to bronze. When nut crops are abundant, the trees are alive with birds and squirrels and—in regions where they still roam—wild turkeys.

American Beech develops many low branches from its upright single trunk. It is very shade-tolerant and grows in a variety of soils, but it requires relatively high amounts of soil moisture.

One could hardly recommend American Beech for planting on small properties, simply because of its great stature, 100 feet tall and 75 feet wide. But where space is adequate, it ranks among the best of the ornamental shade trees. Also, American Beech is unlikely to be affected by pests and diseases.

habitat . . .
FOREST. upland mesic, terraces, ravines, coves, north and east facing slopes
- zone—3b

form . . .
CONICAL. ovoid, large canopy tree (75–100′)
- branching—horizontal, long, spreading, DROOPING branches
- twig—ZIG-ZAG, silver-gray, slender with glossy, LANCE-shaped (3/4″) buds; brown with many scales
- bark—smooth, silver-SLATE, gray with clear straight trunk

foliage . . .
ALTERNATE. simple, oblong-lanceolate, blue-green leaves (2–6″) with coarsely SAW-toothed margin; LEATHERY, yellow-green below with straight, PARALLEL side veins
- color (fall) YELLOW-brown
- season—deciduous

flower . . .
CLUSTER. (male) yellowish green, globular, (3/4–1″) dia. on slender (2″) stalk; (female) reddish (1/4″), hairy scales in pairs
- sex—monoecious, with new unfolding leaves

fruit . . .
NUT. tan-brown, 3-ANGLED (5/8″), enclosed in woody, short-stalked, 4-valve, BUR-like (1″) husk; prickly
- season—maturing in autumn, splitting open

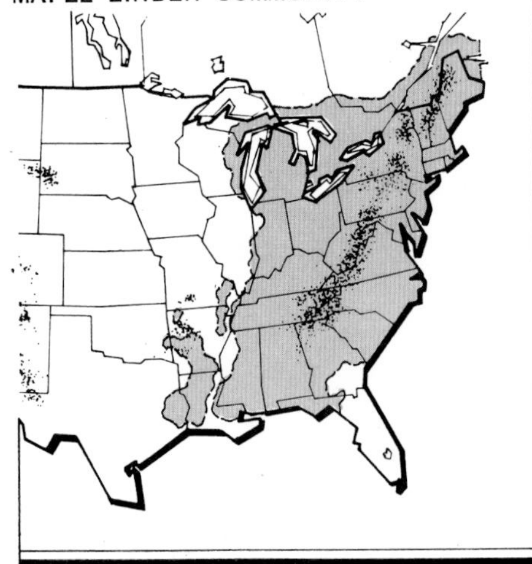

Fagus grandifolia · American Beech

White Ash, the most common ash, is a handsome tree for lawns, streets, and parks. It grows vigorously in a variety of soils but prefers those that are deep and moist. It is common to cool, shaded slopes.

White Ash grows quickly when young (up to 20 feet in 10 years), and it lives for a moderately long time. Its winged seeds are the preferred food for many song and game birds. White Ash is also one of the first trees to change color in the fall. Its fall color is very striking, ranging from yellow to deep shades of purple.

The early pioneers found White Ash very useful for supplying their building and heating needs and for countless small, everyday uses. The strong, clean-grained, enduring wood served many practical needs and even today vies with oak for the honor of being the most valuable American timber tree.

White Ash has become increasingly popular in recent years, and it is now available from most nurseries.

habitat . . .
FOREST. upland mesic, mesic-dry, ravines and upland slopes
- zone—3b

form . . .
IRREGULAR. pyramidal, large canopy tree (75–100′)
- branching—dense, RECURVING, coarse branches
- twig—smooth, yellow-green, stout with leaf scar nearly SURROUNDING glabrous bud; dark-brown
- bark—gray, thick, coarse-fissured with ridges of DIAMOND-like pattern

foliage . . .
OPPOSITE. pinnately COMPOUND, (5–7) leaflets, oblong-lanceolate, bright green leaves (2½–5″) with entire margin; smooth, pale whitish below
- color (fall) yellow-molting PURPLE
- season—deciduous

flower . . .
CLUSTERS. (male) purplish, compact (¼″), dense clusters; (female) petalless, loose clusters
- sex—dioecious

fruit . . .
SAMARA. tan-brown, winged (1–2″) KEY in dense, drooping clusters
- season—maturing late summer, early autumn

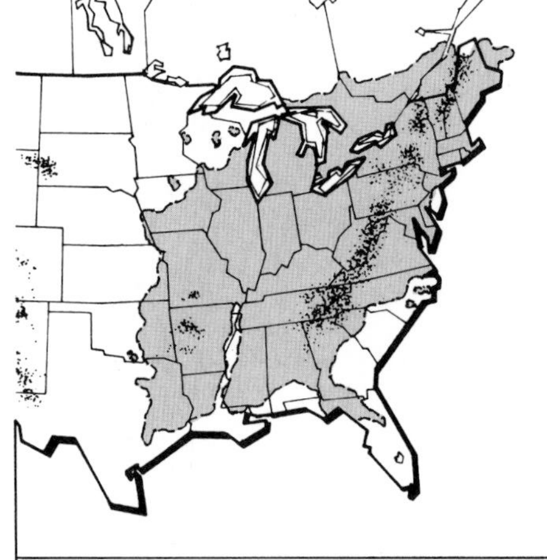

Fraxinus americana · White Ash

Tuliptree is the tallest hardwood tree in North America. It grows rapidly and often reaches heights of over 150 feet in the woodlands, where it does well competing with other species for sunlight. Under such conditions, the vertical trunk may not have a single side limb for its first 50 feet. Tuliptree is essentially a southern tree, but it is found north to New England and west to Indiana and Illinois.

The name, Tuliptree, comes from the shape of its leaf and from its large magnolia-like flowers. It is also referred to as Yellow Poplar, in reference to its soft wood and poplar-like ridged bark. In fact, the tree is neither a tulip nor a poplar but a member of the Magnolia Family, as indicated by its flower.

The flower is green and yellow and blooms when the leaves appear at the tips of branches, which are often many feet above the ground. Because of this, the flower is seldom seen up close. As the photograph shows, the search for a rare, low-hanging branch is worth the effort. The flowers mature into straw-colored, cone-shaped seedheads. These seedheads persist well after the leaves have fallen, scattering their seeds to provide next year's seedlings.

Tuliptree's resistance to smoke makes this tree a good choice for street plantings. It grows rapidly on good soils and can live for 200 years.

habitat . . .
FOREST. upland mesic-dry, low slopes, hills, sheltered coves and stream valleys
- zone—5a

form . . .
COLUMNAR. pyramidal, large canopy tree (75–100′)
- branching—irregular, ASCENDING limbs
- twig—stout, glossy red-brown, slender, zig-zag, encircled by leaf SCAR with flattened, DUCKBILL-like, velvety buds; greenish red-brown
- bark—gray with orange-green fissures, shallow-furrowed

foliage . . .
ALTERNATE. simple, palmately (4–6) LOBED, glossy bright green, TULIP-like leaves, (3–6″) across with entire margin; pale beneath
- color (fall) YELLOW
- season—deciduous

flower . . .
SINGULAR. upright, yellow-green, CUP-shaped (2–3″) across with orange-tinged petal edges; FRAGRANT
- sex—dioecious

fruit . . .
SAMARA. woody, tan-brown PYRAMIDAL (1–2″) spike with 1 or 2 seeded nutlets; overlapping
- season—maturing in autumn, PERSISTING through winter

MAPLE-LINDEN COMMUNITY

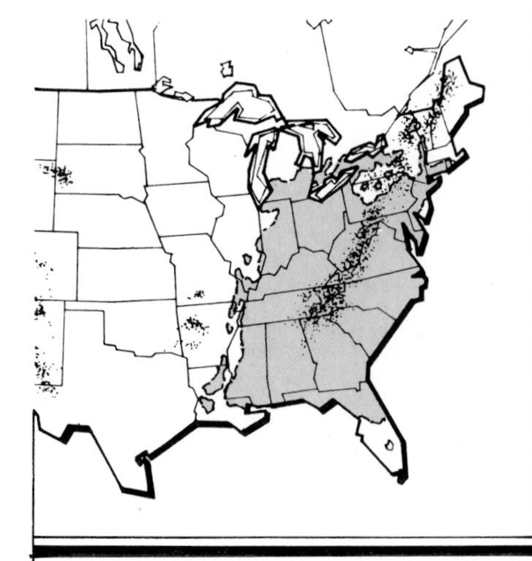

Liriodendron tulipifera · Tuliptree

Northern Red Oak is the only oak in the Maple-Linden community. No other oak can germinate and grow in this plant community's dense shade. Northern Red Oak develops a sturdy trunk with dark, deeply furrowed bark and high branches that reach upward. When grown in the open, it develops a symmetrical crown. Northern Red Oak is a very hardy tree that can live for 200 to 300 years. It is easy to transplant and is one of the fastest growing oaks.

The term "red" in this oak's name clearly refers to its leaves. In early spring, the leaves emerge as a deep red. In autumn, they become dark maroon-red to brown. Northern Red Oak has lobed and bristle-tipped leaves, like other members of the Black or Red Oak Family. But leaf form can vary significantly even on one tree; Northern Red Oak's leaves are generally similar to those of the Black, Pin, Northern Pin, and Scarlet oaks. Thus, leaves alone are an unreliable means of plant identification.

The acorn provides the surest method of identification where oaks are concerned. The size and shape of the acorn and cup are the best guides to identifying the tree that produced it. The Northern Red Oak acorn is fat, round, and large in comparison to other acorns. Its cup is shallow, covering at most a quarter of the acorn.

Northern Red Oak is an excellent shade tree as well as a valuable timber tree. It is unique among northern oak species for other reasons as well: its range extends farther north, and it is usually the tallest.

habitat . . .
FOREST. upland mesic, mesic-dry, coves, ravines, and north and east facing slopes
- zone—3a

form . . .
OVOID. globular, large canopy tree (75–100′)
- branching—stout, ASCENDING limbs
- twig—slender, glossy red-brown with smooth, sharp-pointed, clustered end buds
- bark—dark gray-black, fissured base with shallow, wide-FLAT, gray plates

foliage . . .
ALTERNATE. simple, pinnately (7–11) LOBED, dark green leaves (4–9″) with shallow, BRISTLE tipped sinuses; base wedge-shaped, glabrous beneath
- color (fall) yellow-BROWN
- season—deciduous

flower . . .
CATKIN. slender, yellowish green (3–4″), drooping clusters
- sex—monoecious

fruit . . .
ACORN. brown, egg-shaped (5/8–1 1/8″), 1/3 enclosed by shallow cap; scales tightly overlapping, PAPERY red-brown
- season—maturing second year

MAPLE-LINDEN COMMUNITY

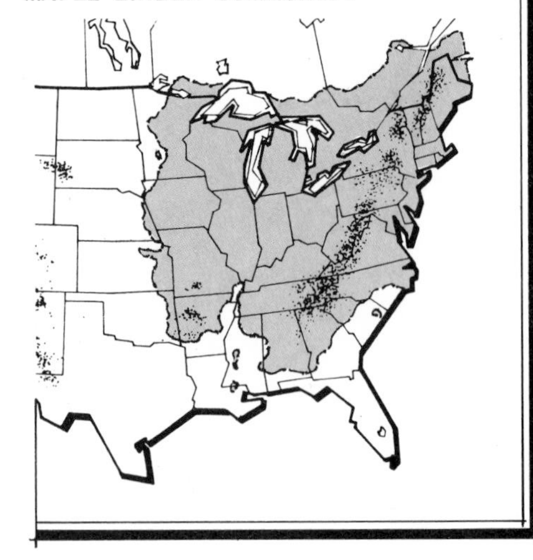

Quercus borealis · Northern Red Oak

American Linden, often referred to as Basswood, is usually a tall, round-topped tree found on rich, moist soils in the company of hard maples. Linden tends to sprout at the base of the trunk and later produces a clump of tall trees around what was the original tree.

Linden flowers bloom in June and July. They are small and relatively inconspicuous, except for their fragrance, which is very strong and may carry for several blocks. Linden ranks with Black Locust and Russianolive in this olfactory expression.

The small flower clusters, with their bract-like, winged stems, practically drip with nectar, as honeybees come from far and near. For days, the air is vibrant with their humming. When the fruit develops into gray, pea-sized seeds, the winged stems propel them away from the tree, like small helicopters moving with the wind.

Linden's wood is white, close-grained, free of knots, and easily worked. For years, it has been a woodcarver's delight. It is also used for furniture, charcoal, and house trim.

Linden is typically a stately and sturdy tree with massive main limbs. As a specimen, it develops a narrow, symmetrical form and normally grows to 75 feet in height. Occasionally it grows to a height of 125 feet. In summer, no other tree has denser shade; at the same time, Linden has an open appearance. American Linden grows at a medium-fast rate (30 feet in 20 years) and commonly lives 150 to 200 years.

Tilia americana · *American Linden*

habitat . . .
FOREST. upland mesic, coves, mesic ravines, and north and east facing slopes
- zone – 3a

form . . .
OVOID. dense, large canopy (75–100′)
- branching – ascending, symmetrical, often DROOPING limbs
- twig – slender, reddish-gray, slightly ZIG-ZAG with glossy, 2-scaled bud; RED
- bark – gray-brown, smooth, becoming furrowed with long narrow ridges; often MULTIPLE trunk

foliage . . .
ALTERNATE. simple, cordate, HEART-shaped base, dark green leaves (4–8″) often as wide with COARSE-toothed margin; palmately veined, light green beneath
- color (fall) YELLOW-gold
- season – deciduous

flower . . .
CLUSTER. yellowish-white (3/8–1/2″), FLAT-topped cluster on leafy BRACT-like wing; fragrant
- sex – monoecious

fruit . . .
SAMARA. globular, tan-brown (3/8″) pairs (2–3) with fine, gray hairs; nutlet of WOODY balls on bract-like wing
- season – maturing late summer, often PERSISTING through winter

MAPLE-LINDEN COMMUNITY

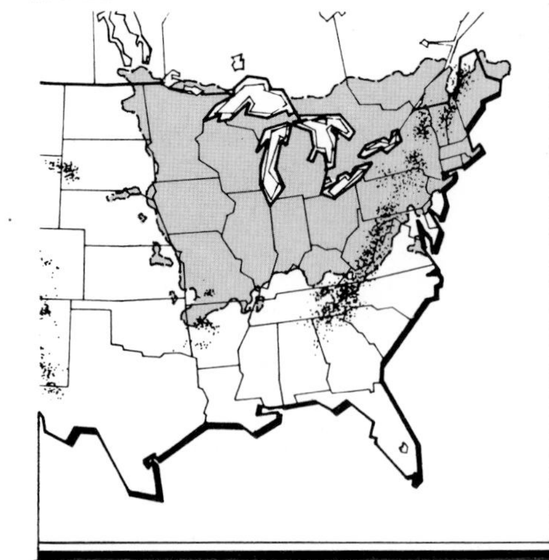

Canada Hemlock, or Eastern Hemlock, is a handsome evergreen tree. It is straight of trunk, long of branch, and majestic in maturity. It is tolerant of dense shade, and it commonly grows along ravines and rocky ridges where there are cool, moist slopes.

Its short-needled, feathery branches make it one of the best ornamental evergreens, as a specimen tree and as an 8-foot tall hedge that can be restrained for years with regular pruning. Canada Hemlock looks beautiful as a tall background plant or in a group in a far corner where a dominant accent is required all year. Where space is available, a Hemlock grove can be planted.

If the soil and temperature are agreeable, Canada Hemlock can be a fast grower. However, most authorities classify it as a slow-growing tree that lives for many years.

habitat . . .
FOREST. upland mesic, sheltered coves, ravines, north and east facing slopes
- zone—3b

form . . .
CONICAL. dense, large canopy tree (75–100′)
- branching—horizontal, SPREADING, often drooping branches
- twig—slender, yellow-brown, rough, PEG-like base; pendulous ends with brownish, blunt buds
- bark—SCALY, red-brown; becoming deep-furrowed with flat-topped ridges

foliage . . .
LINEAR. flattened, appearing 2-ranked, dark green needles (3/8–3/4″) with short leaf stalk; fine-textured with 2 WHITISH bands beneath
- color (fall) GREEN
- season—evergreen

flower . . .
CONE. (male) yellow, (1/8″) dia. clusters; (female) green, singular
- sex—monoecious

fruit . . .
CONE. small, tan-brown, elliptic (1/2″) with smooth-margined scales; PENDULOUS at ends of twigs
- season—maturing late autumn

MAPLE-LINDEN COMMUNITY

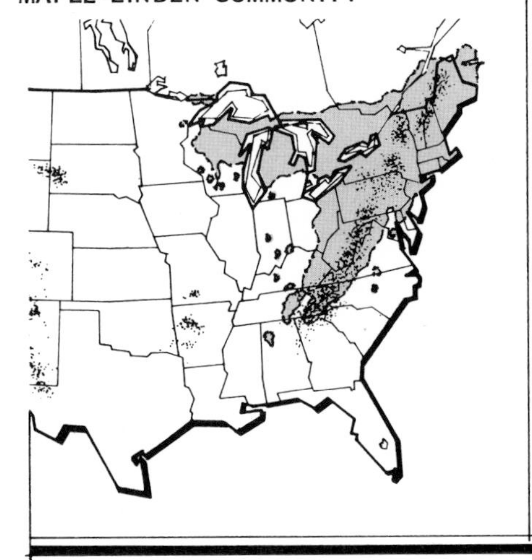

Tsuga canadensis · Canada Hemlock

Shadblow Serviceberry is a common, small understory tree or large shrub that is found in cool, moist woodlands. Early settlers named it Shadblow or Shadbush, because it blooms about the time the shad run upstream from the sea to spawn. The name, Serviceberry, may have come from the southern mountain region. There, it bloomed each year at about the same time that the preacher "went to remote hollows to hold religious services for those who had died, had been born, or had taken up living together since the preacher's last trip" (Borland, 1983). At any rate, Serviceberry has been much admired since colonial times and for good reason.

Serviceberry usually grows in clumps with many upright branches. This small tree, with its delicate, open crown, blooms early in spring with many white, long-petaled blossoms. In midsummer, the small, crimson-colored, apple-like fruit develops. Serviceberry is equally appealing in fall, when the foliage changes to shades of deep orange and rusty red. Serviceberry is not only an attractive tree but also has tasty fruit. Early settlers used Serviceberries in a mixture of dried fruits and meat called "pemican." It feeds many species of birds as well.

Shadblow Serviceberry is very hardy. It is an excellent choice for urban designs and home plantings.

Amelanchier canadensis · Shadblow Serviceberry

habitat . . .
FOREST. upland mesic, north and east facing slopes
- zone – 3a

form . . .
GLOBULAR. slightly leggy, large understory tree (35–50′)
- branching – slender ASCENDING branches
- twig – slender, gray, slightly exfoliating with narrow, long-POINTED, chestnut brown buds; scales with silvery hairs
- bark – smooth, SILVERY-gray with slightly twisting trunks

foliage . . .
ALTERNATE. simple, elliptic-oval, dark green leaves (1½–2″) with FINE-toothed margin; apex pointed with sharp tip, white-hairy (young) dull, paler beneath
- color (fall) red-ORANGE
- season – deciduous

flower . . .
SPIKE. erect, white (⅛–⅜″) pyramidal (1–3″) spikes
- sex – monoecious, before leaves appear

fruit . . .
BERRY. globular, RED/maroon-purple (½″), dia., drooping clusters
- season – maturing early summer

MAPLE-LINDEN COMMUNITY

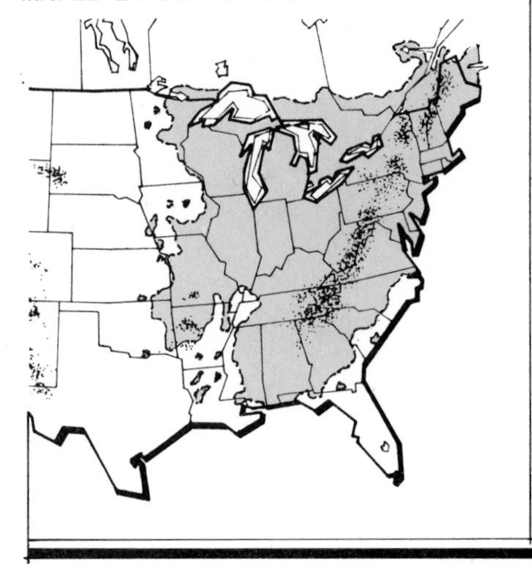

American Hornbeam is a small understory tree, usually less than 20 feet tall. It commonly grows in cool, moist woodlands and tolerates many soil types and sun or shade.

American Hornbeam is an attractive little tree that has several distinctive characteristics. One is its smooth, slate-gray bark on fluted trunks that resemble muscles. This feature has given rise to common names such as Blue Beech and Musclewood.

The male and female flowers appear in early spring in separate catkins on the same tree. The fruit that develops is unusual. It is a tiny, ribbed nut, held in clusters of many small, three-lobed, leaf-like bracts. In fall, the foliage turns yellow, orange, red, or tan.

The wood of American Hornbeam is extremely hard and tough, as its name implies. In the name hornbeam, "horn" represents toughness and "beam" is a term for tree. American Hornbeam grows very slowly, often taking up to 150 years to reach its mature height.

habitat . . .
FOREST. upland mesic, lowland wet-mesic, cool north and east facing slopes and lower slopes
- zone—2

form . . .
GLOBULAR. dense, large understory tree (35–50′)
- branching—horizontal, SPREADING, zig-zag limbs on multiple trunk
- twig—slender, red-brown, slightly ZIG-ZAG with scaly (1/8″), rust-brown buds; WHITE margin scales
- bark—smooth, thin BLUE-gray; fluted MUSCLE-like appearance

foliage . . .
ALTERNATE. simple, elliptic-oval, dark blue-green leaves (2–5″) with DOUBLE-toothed margin; veins to leaf margin, pale beneath
- color (fall) SCARLET-orange
- season—deciduous

flower . . .
CATKIN. (male) green (1½–2″), drooping clusters; (female) reddish green (½–¾″), paired catkins
- sex—monoecious

fruit . . .
SAMARA. red-orange, 3-LOBED, (2–4″) winged clusters; pendulous
- season—maturing late summer

MAPLE-LINDEN COMMUNITY

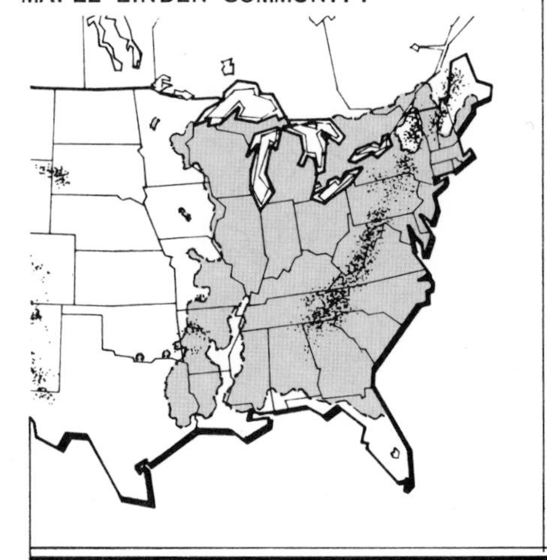

Carpinus caroliniana · American Hornbeam

Pagoda Dogwood, or Alternate-leaved Dogwood, is a handsome, small understory tree. It grows 15 to 20 feet tall and has stately, horizontal branches. It is the only dogwood with an alternate leaf arrangement. The leaves are usually clustered at the tips of its many branchlets.

Pagoda Dogwood commonly grows in the understory layers of cool, moist woodland; on mountain slopes and ravines; and along banks of streams, swamps, and lakes. However, it is seldom in great abundance anywhere, so it is a pleasant surprise to happen upon one regardless of the season.

In winter, this Dogwood's branches are at their best, because their horizontal, pagoda quality is displayed. Winter reveals the red to green color of new twig growth and the strongly horizontal, layered branching pattern that gives the tree its name. In late spring after the leaves have matured, Pagoda Dogwood comes into bloom. The flowers are creamy white in flat-headed clusters, providing a truly memorable sight when highlighted by filtered sunlight.

In late summer and early fall, the fruit develops. As a cluster of blue-black berries on a bright red stem. At this time, birds bring the tree to life, and only a few berries are left by the birds to fall to the ground. The leaves in autumn turn deep maroon in shadowy places; in sunlight, tints of scarlet and orange appear in bright profusion.

Cornus alternifolia · Pagoda Dogwood

habitat . . .
FOREST. upland mesic, sheltered coves, floodplain ravines and terraces
- zone—3b

form . . .
OVOID. open, small understory tree (20–35′)
- branching—horizontal, STRATIFIED branches
- twig—slender, green, reddish brown, lustrous, stout twigs encircled by leaf scar with small pointed buds; red-brown
- bark—smooth, green-brown, slightly fissured; often multiple trunk

foliage . . .
ALTERNATE. simple, oblong-ovate, bright green, long-pointed leaves (2½–5″) with entire margin; veins curved upward along margin, whitish beneath
- color (fall) yellow, MAROON-red
- season—deciduous

flower . . .
CLUSTER. creamy white, (2″) wide, FLAT-TOPPED, erect clusters; terminal
- sex—monoecious, fragrant

fruit . . .
BERRY. globular, bluish, RED-purple, (¼″) dia., flat-topped (3″) across clusters on red stems
- season—maturing late summer

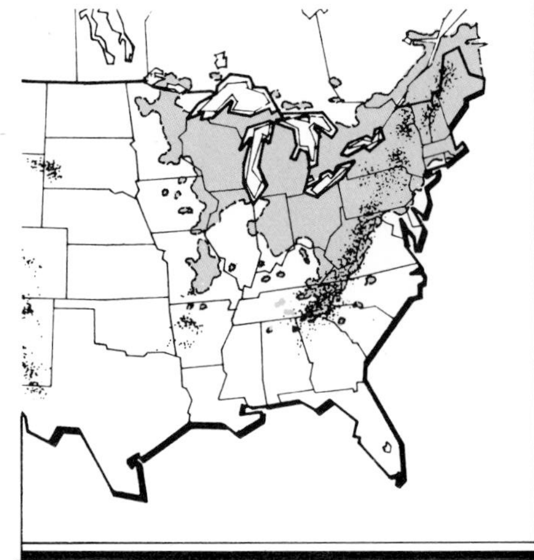

Common Witchhazel usually grows as a shrub, 10 to 12 feet high, but it may become a scraggly tree as tall as 25 feet. It commonly grows at the edges of woods and fields on any fertile soil, either moist or dry. It prefers a sunny location or partial shade.

A unique feature of Common Witchhazel is that it blooms very late in the fall, sometimes even after snowfall, hence its common name Winter Bloom. Witchhazel's blooms look like a tangle of narrow, bright yellow ribbons, but they are actually only four petals that twist and curve so they look like a tassel.

Common Witchhazel is also unique in its seed dispersal. Just before flowering, the nuts dry and mechanically snap open, flinging their nutlets some 15 to 20 feet from the plant. This gives rise to another common name, Snapping Alder. The leaves are also very attractive, turning golden yellow with the blossoms in late fall, making it a welcome ornamental for a shrub border.

An extract is obtained from Witchhazel bark and twigs that is available in drug stores for use on mosquito bites.

habitat . . .
FOREST. upland mesic, ravine slopes, bottoms, and mesic open, woods
- zone—5a

form . . .
IRREGULAR. globular, open, small understory tree (20–35′)
- branching—HORIZONTAL, wide-spreading branches, often on leaning trunks
- twig—slender, yellow-brown, ZIG-ZAG with velvety gray hairs having ovate, clustered (½″) buds; stalked orange-brown
- bark—smooth, gray-brown, narrow, somewhat scaly ridges with HORIZONTAL lenticels

foliage . . .
ALTERNATE. simple, oblong-OBOVATE, bright green leaves (3–5′) with WAVY upper half margin; base uneven, dull, slightly tomentose
- color (fall) YELLOW
- season—deciduous

flower . . .
CLUSTER. bright yellow, (4) linear (½–1″) wide, THREAD-like petals, short-stalked; TWISTING
- sex—monoecious

fruit . . .
CAPSULE. tan-brown, two-BEAKED, (½–¾″) WOODY splitting capsule; seeds (¼″) black
- season—maturing in autumn

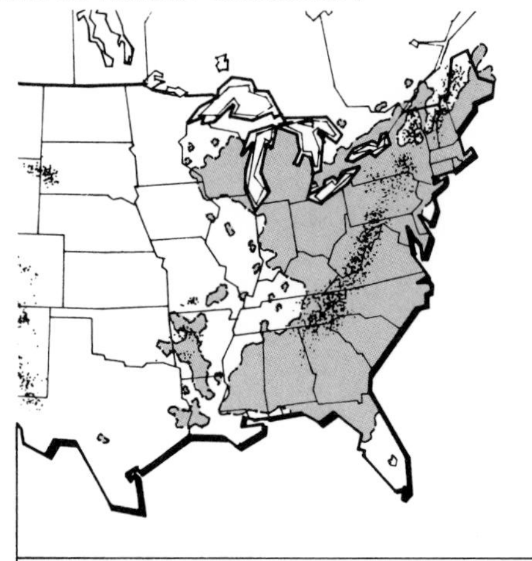

Hamamelis virginiana · Common Witchhazel

Atlantic Leatherwood is a small shrub, 2 to 6 feet in height. It grows in moist, rich soils at the base of mesic slopes or in cool, shaded swales of north- and east-facing wooded slopes.

Leatherwood blooms early, and the flowers are small and pale yellow. The fruit quickly matures into small, red, egg-shaped clusters. The fruit is rarely seen because it is hidden among the leaves; then it matures and drops early.

The most unusual characteristic of Atlantic Leatherwood is its bark. While its outward appearance is unassuming, its toughness is astonishing. The wood is easily broken, offering no great resistance. But the bark is another matter. It neither yields nor gives. It can be bent or twisted, but it simply does not split. Indians understood its properties and used this plant for bowstrings and fish lines (Keeler, 1903).

The decorative value of Atlantic Leatherwood lies not in its flowers or fruit but in its light yellow-green foliage. This shrub can be used to best advantage to brighten a dark, shady place in a secluded corner of the garden.

habitat . . .
FOREST. lowland wet-mesic, mesic, moist wood, ravines, and north and east facing slopes
- zone—3a

form . . .
GLOBULAR. symmetrical small shrub (3–8′)
- branching—stout, flexible with upward, curving limbs
- twig—smooth, yellowish brown, SWOLLEN nodes appearing SOCKET-jointed by leaf scar with small DOME-shaped, woolly buds; brown
- bark—smooth, gray-brown; flexible with tough interlacing fibers

foliage . . .
ALTERNATE. simple, elliptic-obovate, glossy, green leaves (2–3″) with entire margin; base wedge-shaped with short petiole, whitish-downy beneath
- color (fall) YELLOW-brown
- season—deciduous

flower . . .
TUBULAR. small, yellowish white (1/4–3/8″), BELL-shaped clusters of 3 and 4; short-stalked
- sex—monoecious, appearing before leaves

fruit . . .
BERRY. small, reddish yellow-green (3/8″), EGG-shaped, dense clusters; often inconspicuous
- season—maturing early summer

MAPLE-LINDEN COMMUNITY

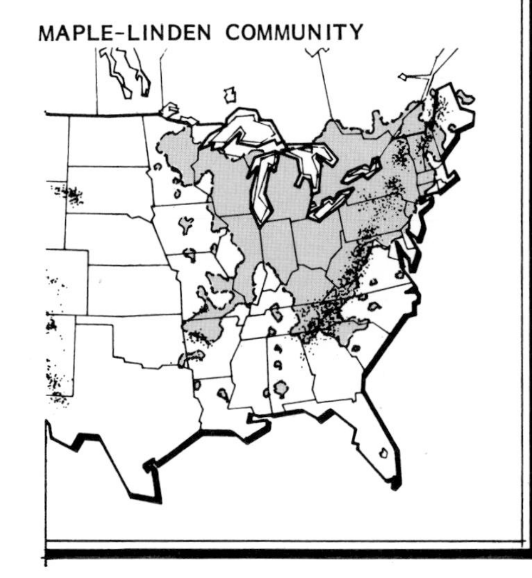

Dirca palustris · Atlantic Leatherwood

Common Snowberry is used more extensively for ornamental plantings than most native shrubs. In its wild state, it is found on dry, rocky banks, talus slopes, or shady cliffs; in the garden, it grows just about anywhere in sun or shade.

In spring, Common Snowberry develops leaves and is simply a clean, bright little shrub. In July, it begins putting forth flower clusters of tiny pink bells, and this continues until September. These flowers are so small that they are seldom noticed. It is when these flowers transform into clusters of snowy white balls, varying in size from small peas to small marbles, that they begin to draw attention to themselves. This shrub, laden with its white burden and bent nearly to the ground, is a wonderful sight to see. Snowberry holds its berries intact until the frosts of November.

Besides being a desirable ornamental, the low, dense form of Common Snowberry makes it an excellent shrub to control soil erosion and for wildlife cover.

habitat . . .
FOREST. upland mesic, mesic-dry, dry, open rocky slopes, cliffs, wooded hillsides and dry clay banks
- zone—2

form . . .
GLOBULAR. ovoid, leggy open shrub (2–5′)
- branching—slender, erect, ARCHING limbs
- twig—slender, brown, smooth to wiry with small brown buds; minute
- bark—hollow, gray stems

foliage . . .
OPPOSITE. simple, elliptic-oblong, bright green leaves (¾–2″) with entire or sinuately lobed margin; apex rounded, hairy veins beneath
- color (fall) yellow-BROWN
- season—deciduous

flower . . .
CLUSTER. small, pinkish white (⅛″), BELL-shaped cluster along stem; terminal or nearly so
- sex—monoecious

fruit . . .
BERRY. white (¼″), dense, drooping clusters; TERMINAL
- season—maturing late summer, PERSISTING through winter

MAPLE-LINDEN COMMUNITY

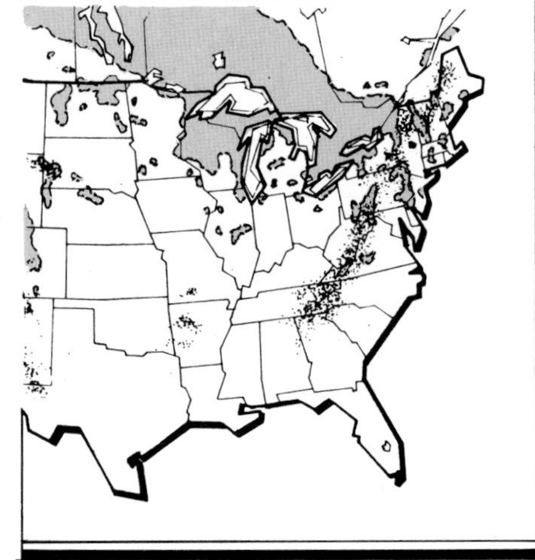

Symphoricarpos alba · Common Snowberry

One of the most delightful plant communities is not really a community at all. Instead, this group of plants provides ground cover for several communities. These plants are found in the Oak–Hickory and Mixed Floodplain communities, but they also flourish in the Maple–Linden community.

Spring woodland flowers are remembered with great affection because they bloom for such a short period of time. Some people describe them as the flowers that can't wait for spring, and appropriately so. Most of them bloom early, before the dense woodland leaf canopy blots out the sun. They manage to emerge, bloom, seed, and wither in only a few short weeks. But the memory of woodland blooms stays throughout the year.

A few, like Canada Wildginger, do not wither but remain as a groundcover throughout the summer. The Sharplobe Hepatica is not only one of the showiest of the early woodland flowers, but it defies all rules and holds its red-brown leaves under the snow throughout winter until blooming time next April. But these are exceptions in the elite order of spring ephemerals.

Many nurseries and seed catalogues now sell spring wildflowers for use in the shaded garden. It is recommended that these delicate perennials not be taken from the woods but purchased from those who grow them commercially.

Woodland Flower Community

American Maidenhair Fern is one of the most delicate and attractive ferns. It is most abundant in the rich, shaded soils of ravines and beneath moist, rocky banks of limestone outcroppings.

The leaf fronds are circular or shaped like a horseshoe, giving the appearance of a crown of laurel leaves to be placed in a maiden's hair.

The leaves, which are 6 inches or more above the ground, appear to be floating mysteriously; their small, wiry stems are hardly noticeable.

habitat . . .
WOODS. moist, rich woodlands, ravines, rocky banks, limestone cliffs and shaded slopes
•zone—3b

form . . .
ERECT. slender, circular, flat, HORSESHOE-shaped, stemmed fern (8–20″)
•rhizome—gray-brown, thick with light brown scales at tip; CREEPING
•leafstock—shiny reddish brown-black, long, slender, smooth, BRITTLE
•roots—slender, gray at ends of rootstock
•fiddleheads—wine red, multiple in form

foliage . . .
FRONDS. pinnatifid COMPOUND, HORSESHOE-arranged leaves (5–6), delicate gray-green LACY leaves (8–16″) and as wide with subleaflets, flat, FAN-shaped (10–26) pairs; alternately arranged, lower edge entire, short-stalked with veins free
•season—semi-evergreen

SPORES. small, rounded, brown (1–5) fruitdots along upper margin of leaflets; indusium (curling of leaf), white to yellowish green, thin, linear in shape
•season—maturing late summer

WOODLAND FLOWERS

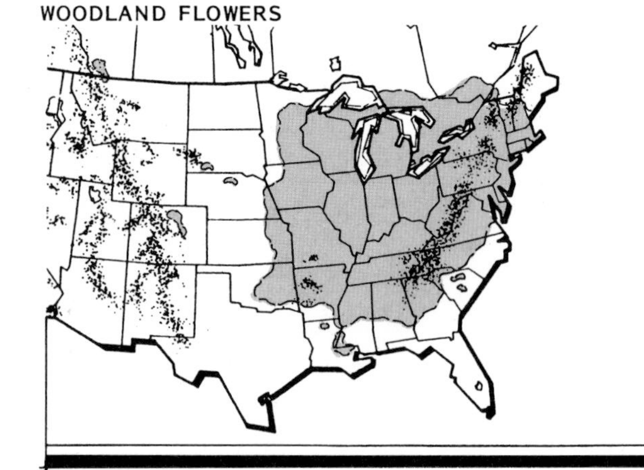

Adiantum pedatum · American Maindenhair Fern

Wild Leek is usually found in colonies on the moist, rich soils of woodland slopes and bottomlands.
Its wide leaves rise out of an onion-shaped bulb and wither before the flowers appear in June or July. The white flowers grow on a slender stem. The fruit that follows looks like a halo of shiny, black, ball-shaped clusters.

habitat . . .
WOODS. moist, rich wooded coves
- zone—3a

form . . .
ERECT. glabrous, leafless scape herb (6–20″)
- perennial—bulb with slightly fibrous coat (ONION-like)

foliage . . .
OPPOSITE. simple (2–3) linear blades, flat to wide elliptic leaves (8–12″) with entire margin; basal leaf sheath, WITHERING before flower appears

flower . . .
SOLITARY. creamy white (¼″), DOMED clusters, (1½″) dia. on slender (4–12″) scape; ERECT
- season—early spring (April–July)
- sex—monoecious, FRAGRANT

fruit . . .
CAPSULE. small, (⅛″) globular, 3-LOBED, ball-like clusters; SHINY black, pebbled seeds resembling buckshot
- season—maturing late summer

WOODLAND FLOWERS

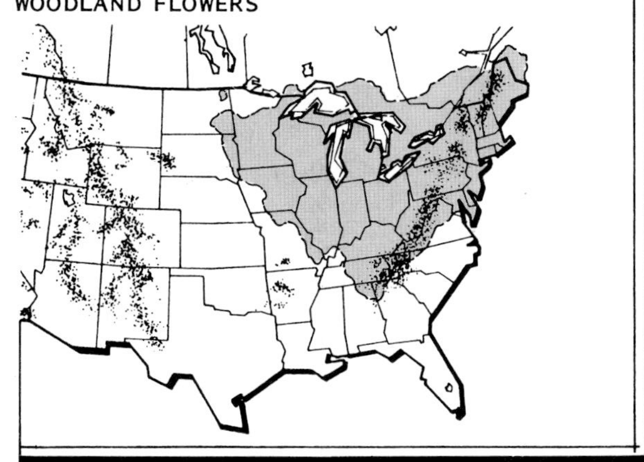

Allium tricoccum Wild Leek

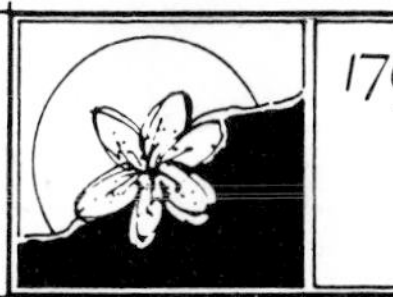

American Wood Anemone grows in colonies in the rich soils of open woodlands and prairies. It blooms from April to June and is the earliest and smallest woodland anemone. Unlike Rue and False Rue Anemone, Wood Anemone blooms with a solitary flower rather than with several flowers clustered on a stem.

Mesquakie Indians made tea from anemone roots to treat headaches and dizziness (Runkel and Bull, 1979).

habitat . . .
CLEARINGS. moist, rich open woods and rocky woodland border
•zone – 3a

form . . .
ERECT. low, slender-stemmed herb (4–8″)
•perennial – underground tuber (corms)

foliage . . .
OPPOSITE. whorled, palmately COMPOUND, leaflets (3–5), bright green-maroon leaves (1¼″) with sharp-toothed segments; lobes rounded, glabrous

flower . . .
SOLITARY. white (1–1¼″) with petals absent; often pink on backside
•season – early spring (March–May)

fruit . . .
ACHENE. brown, seed-like, hairy cluster
•season – maturing early summer

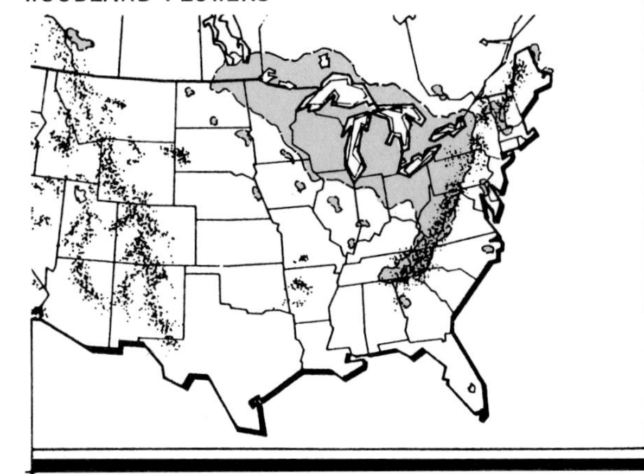

Anemone quinquefolia · American Wood Anemone

Rue Anemone is found in early spring on sloping areas of dry, open woods. It has flowers in clusters of two or three.
The flowers of Rue Anemone last longer than those of other early spring ephemerals.

The underground tubers of this plant were eaten by early pioneers and Indians. Today, however, Rue Anemone is becoming uncommon and should be left undisturbed, as it will not tolerate root division (Runkel and Bull, 1979).

habitat . . .
WOODS. moist, rich woodlands
- zone – 4b

form . . .
ERECT. small, glabrous, delicate-stemmed herb (4–8″)
- perennial – tuberous roots

foliage . . .
OPPOSITE. ternately COMPOUND, leaflets (3–5) ovate to suborbicular, bright green leaves (½–1″) with entire 3-LOBED margin; somewhat whorled

flower . . .
DISK. white to pink, singular, (1″) dia. with (5–10) obovate sepals
- season – early spring (April–June)
- sex – dioecious

fruit . . .
ACHENE. prominently ribbed (8–10), one-seeded
- season – maturing early summer

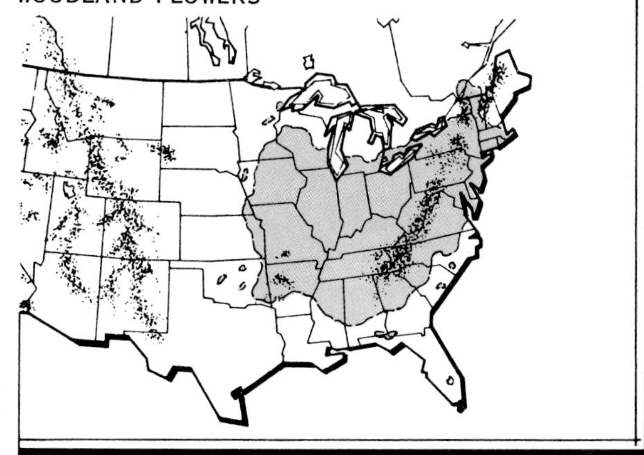

Anemonella thalictroides · Rue Anemone

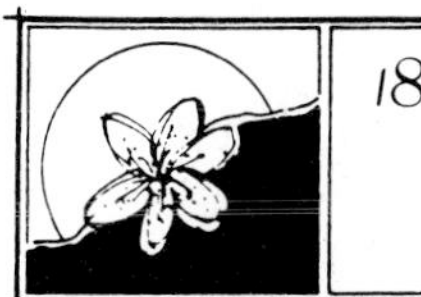

American Columbine is at home in a wide variety of conditions, from dry soil on sunny cliffs to shady, moist soil in woodlands. It blooms from May to July and has drooping red and yellow bell-shaped flowers.

The Columbine flower has distinct, narrow rear tubes or spurs that allow access to its nectar glands. Only such pollinators as hummingbirds and long-tongued insects like moths and butterflies are allowed access. Bees attempting to reach the nectar by cutting through the tube of American Columbine find a bitter juice that discourages their future efforts (Runkel and Bull, 1979).

habitat . . .
WOODS. forest edge, rocky woodlands, open fields, roadsides and shaded slopes
•zone—3a

form . . .
ERECT. somewhat branched, stout-stemmed herb (1–3′)
•perennial—short rootstocks

foliage . . .
ALTERNATE. ternately COMPOUND, leaflets (3) dark green (1–1½″) with slight-toothed margin; pointed apex, glabrous above, basal leaves

flower . . .
TUBULAR. deep red, five sepals (1½″), petal-like with (5) hollow SPURS pointing upwards; petals yellow on slender, NODDING stems
•season—late spring (June–July)
•sex—monoecious

fruit . . .
FOLLICLE. tan-brown (¾″), 5-parted with numerous black seeds; angled
•season—maturing mid-summer

WOODLAND FLOWERS

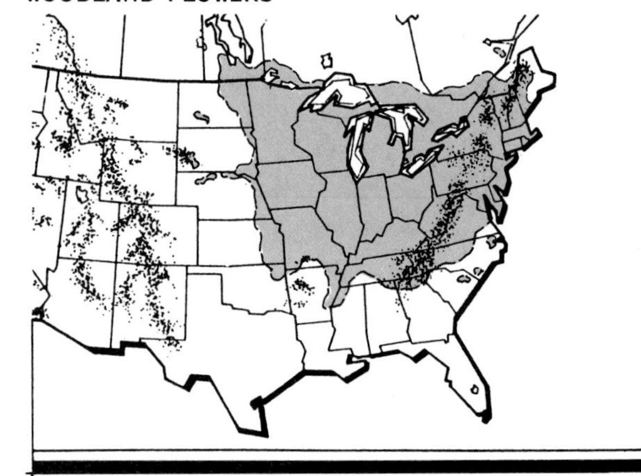

Aquilegia canadensis · American Columbine

Indian JackinthePulpit is an unusual perennial plant that inhabits rich soils of low, moist woods or moist, wooded slopes. It blooms from April to June.

Each plant has one or two leaves, each of which has three large leaflets up to 7 inches long. The flower is made up of a spadix and a spathe. The club-shaped spadix, called the preacher or jack, is covered with tiny yellow flowers. The leaf-like spathe, which is streaked in tones of purple, brown, and green, wraps itself around the lower part of the spadix and curves above it to form the hood or pulpit. When the spathe withers in the autumn, it discloses a cluster of brilliant red berries.

habitat . . .
WOODS. rich, moist woodlands and low alluvial flats
•zone—4a

form . . .
ERECT. slender (2–3) stemmed herb (1½–3′)
•perennial—underground tuber (corms)

foliage . . .
OPPOSITE. pedately COMPOUND, leaflets (3–5) lanceolate-oblanceolate, bright green leaves (6–12″) with round-toothed margin; pinnately divided, 3-LOBED segments

flower . . .
SPIKE. striped, erect, purplish green spadix (2–3″) with spathe (pulpit) sheath-like, CURVING hood
•season—spring, early summer (March–June)
•sex—monoecious

fruit . . .
BERRY. globular, scarlet RED, (⅛″) dia., clustered (2–3″) spike; ERECT, terminating
•season—maturing early autumn

WOODLAND FLOWERS

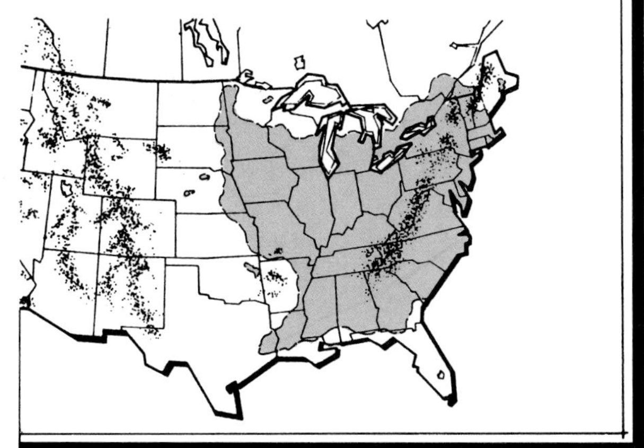

Arisaema triphyllum · Indian Jack in the Pulpit

Canada Wildginger occupies cool, shady woodlands where the soil is moist. It blooms in April and May and has a reddish-brown flower. The flower appears in the notch of two leaf petioles. This flower is so close to the ground that it is sometimes buried by fallen forest litter.

Canada Wildginger is an excellent groundcover for moist, shady places. It has large, heart-shaped leaves that remain green until a killing frost.

Indians and pioneers used this plant to season foods (Runkel and Bull, 1979).

habitat . . .
WOODS. moist, rich woodlands and stream banks
- zone – 3a

form . . .
HERB. low, nearly stemless (6–12″)
- perennial – rhizomatous, somewhat GINGER-like odor; cooked

foliage . . .
OPPOSITE. simple, paired cordate leaves (3–6″) with entire, CREAMY striped margin; glossy bright green (young) with VELVETY appearance

flower . . .
SOLITARY. dark reddish brown-maroon (1½″), CUP-shaped with pointed 3-LOBES at expanded mouth; slender, stalked at base of leaf axils
- season – deciduous, early spring (April–May)
- sex – monoecious

fruit . . .
CAPSULE. fleshy, globular
- season – late spring, early summer

WOODLAND FLOWERS

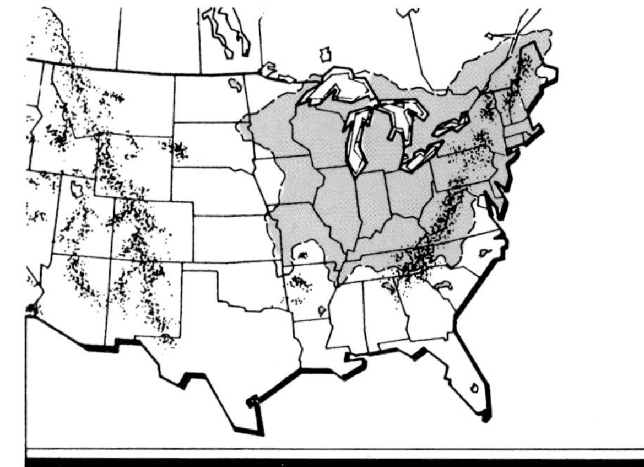

Asarum canadense · Canada Wildginger

Virginia Springbeauty is a spectacular sight when blooming in large patches. It grows in the rich soils of moist, open woodlands. The flowers that bloom from March through May are small. They are white to pink with darker pink veins, and they cluster on the stem. Springbeauty's leaves are narrow and grass-like. The plant disappears in June, and the tubers lie dormant until the next spring.

The starchy tubers of Springbeauty have a sweet chestnut flavor and were eaten as a potato substitute by both Indians and early pioneers (Runkel and Bull, 1979).

habitat . . .
WOODS. moist, rich thickets and clearings
- zone – 2b

form . . .
ERECT. open, slender stemmed, succulent herb (6–12″)
- perennial – underground tuber (corm)

foliage . . .
OPPOSITE. simple (2) linear blades, lanceolate dark green leaves (2–8″) with entire margin; present midway up stem

flower . . .
CLUSTER. small, white to pink (½–¾″), 5-petaled, TINGED with darker pink lines
- season – early spring (Feb.–May), first to bloom, lasting only a few days
- sex – monoecious

fruit . . .
CAPSULE. small, valvate, enclosed by (2) sepals
- season – maturing late spring

WOODLAND FLOWERS

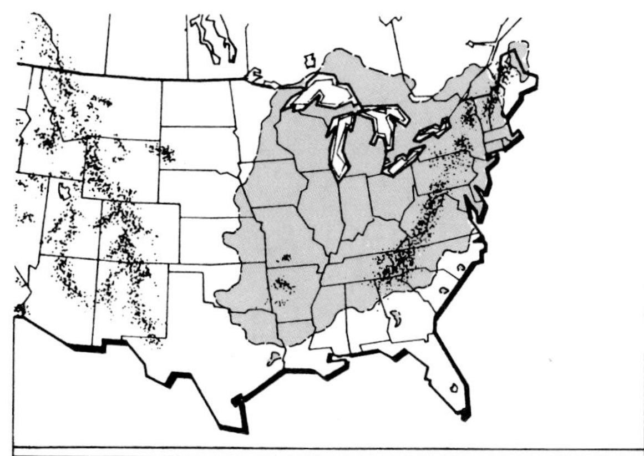

Claytonia virginica · Virginia Springbeauty

Cutleaf Toothwort is also often found in large patches in the rich soils of moist, shady woodlands. The flower of this delicate plant blooms from March to May.

Its name seemingly describes both leaf and flower, but the name Toothwort probably came from the tooth-shaped fleshy tubers. These tubers were gathered in early spring by pioneers, who used them for seasoning many foods (Runkel and Bull, 1979).

habitat . . .
ALLUVIAL. moist, low woodlands and damp thickets
•zone—3b

form . . .
ERECT. unbranched below, singularly stemmed herb (8–16″)
•perennial—TOOTH-like projections, segmented rhizomes; tuberous; rootstock pleasant-tasting

foliage . . .
ALTERNATE. simple, palmately LOBED (3–5) segments oblong-lanceolate, bright green leaves (2–5″) with deeply TOOTHED lobed margin; often appearing in WHORLS of (3) above mid-stem

flower . . .
CLUSTER. small, WHITE to pink (½–¾″) on erect stalk; petals (4) arranged in a CROSS shape
•season—early spring (April–May)

fruit . . .
CAPSULE. slender, flat-angled (1–1½″) pod; opening from bottom
•season—maturing late spring, early summer

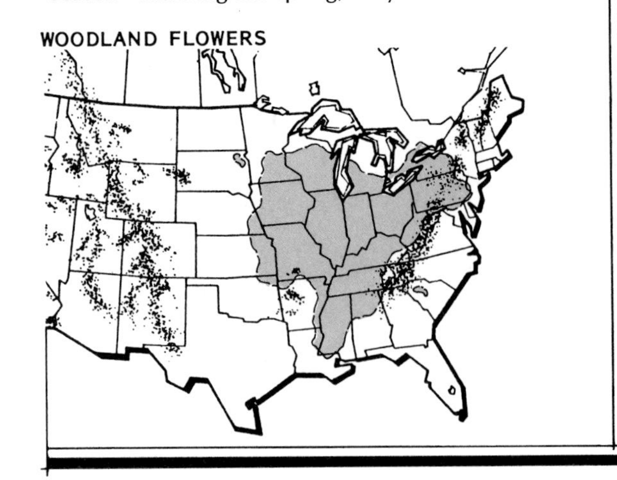

Dentaria laciniata · Cut Leaf Toothwort

Dutchman's Breeches is found in rich soils of moist woodlands, blooming in April and May. The flower is well described by its common name. The breeches hang upside down, attached to the underside of the arching flower stalk. This plant is a close relative of the garden perennial Bleeding Heart.

The roots of Dutchman's Breeches are poisonous, but early pioneers used the plant to treat urinary problems and as a poultice for skin diseases (Runkel and Bull, 1979).

habitat . . .
WOODS. rich, moist woodlands, stream banks and north slopes
•zone – 3a

form . . .
MOUNDING. low, spreading, scapose herb (4–12″)
•perennial – BULBLET-bearing rootstock

foliage . . .
OPPOSITE. pinnately divided, deltoid to ovate, bright green leaves (3–6″) with linear, deep cut entire segments; glabrous, long petioled

flower . . .
SPIKE. fleshy PINK to white (¾″) bulb with pointed scales; diverging SPURS (2) providing the legs of the "Breeches" nodding on arching stem
•season – early spring (April–May)
•sex – monoecious

fruit . . .
CAPSULE. small (1/16″) ellipsoid-linear, opening into (2) parts
•season – maturing early spring

WOODLAND FLOWERS

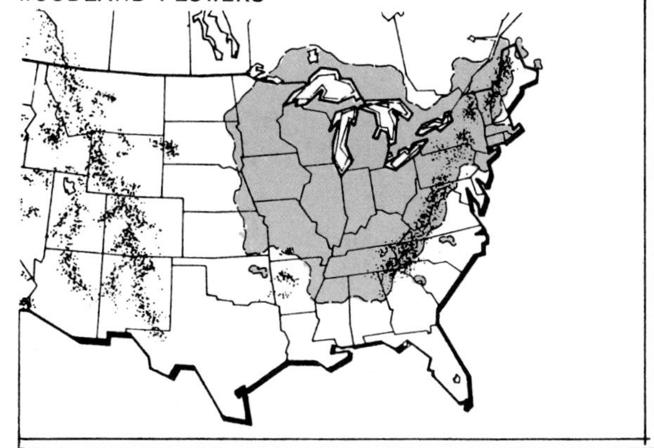

Dicentra cucullaria · Dutchman's Breeches

White Fawnlily has many names; two of the most common are White Troutlily and Dogtooth Violet. It grows in colonies in rich, moist soils on woodland slopes and bottomlands. Some White Fawnlily colonies may be hundreds of years old, and they sometimes cover an area more than 100 feet in diameter.

The leaves of White Fawnlily have a mottled appearance, like a trout or a fawn. The flower and leaves appear between April and June, then they wither and disappear until the following year.

The deeply buried bulb was eaten by Indians and was considered a treatment for gout (Runkel and Bull, 1979).

habitat . . .
WOODS. moist alluvial woodlands, coves and meadows
•zone – 3a

form . . .
ERECT. slender, leafless scape herb (4–10″)
•perennial – glabrous-underground bulb (STOLON)

foliage . . .
OPPOSITE. simple, fleshy, flat elliptic-oblong, brownish green leaves (5–8″) with whitish purple, WATER-like patches; basal sheathing petioles, cylindrical

flower . . .
SOLITARY. tubular white (1″) across NODDING with reddish, separate perianth (6) parted segments; having purplish streaks down midrib-stigma (3)-cleft
•season – early spring (Feb.–May)
•sex – monoecious

fruit . . .
CAPSULE. 3-angled obovoid (½–1″) with CRESCENT-shaped seeds; reddish brown
•season – maturing early spring

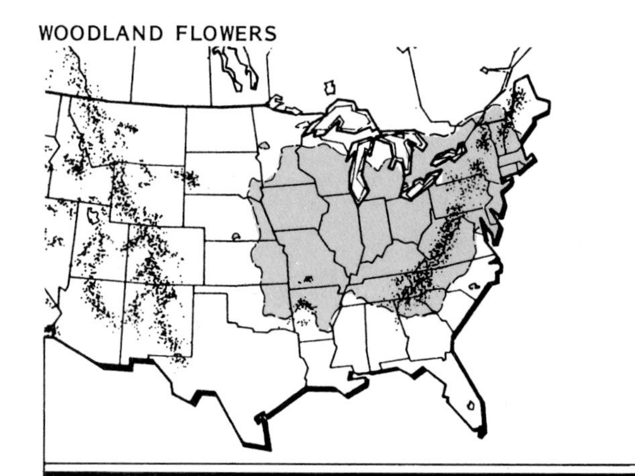

Erythronium albidum · White Fawnlily

Catchweed Bedstraw is an annual groundcover that usually grows in shady woods on damp ground. It blooms in May and June with many small, subtle, white flowers. The stems of this plant are weak, causing it to be prostrate or to recline on other plants for support.

Bedstraw has a novel form of propagation. Its squarish-stems have a rough texture and break easily, allowing segments of the plant to attach to animal fur or clothing. When these segments fall from their temporary host, they root if conditions are right.

habitat . . .
MEADOWS. moist, rich woodlands, open fields, roadsides
•zone – 2b

form . . .
TRAILING. erect to spreading, WEAK scabrous herb (8–30″)
•annual – slender, SQUARE-stemmed, armed with backward-HOOKED bristles

foliage . . .
WHORLED. linear to narrow oblanceolate (6–8) per node, bright green leaves (1–3″) with short, bristly margin; stiff hairs along veins

flower . . .
PERFECT. small, white (1/8″), 4-LOBED clusters in leaf axils; sepals absent
•season – early summer (July–Sept.)

fruit . . .
POD. globular, brown-black (1/8″), dry pod; BRISTLY
•season – maturing early autumn

*Galium triflorum (fragrant bedstraw), a similar species, having smooth stems and leaves mostly in whorls of (6)

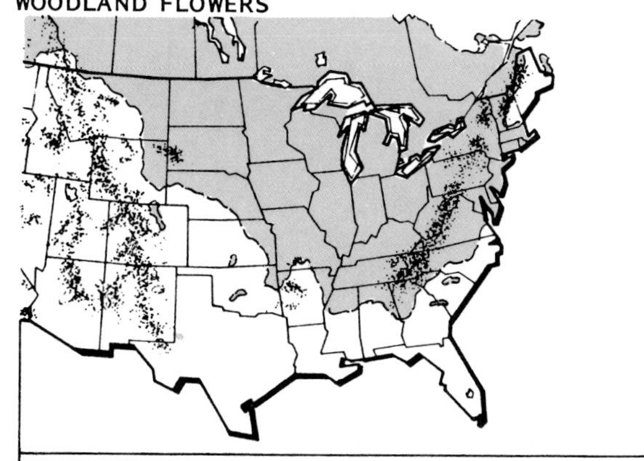

Galium aparine · Catchweed Bedstraw

Spotted Geranium blooms from May to June in rich, moist soils in open woodlands and ditches. It is easily recognized by its palmately lobed leaves and distinctive seed capsules. It is often referred to as Cranes-bill, because of its narrow-beaked seedpods.

Indians used the roots of Spotted Geranium to produce a tea to treat toothaches (Runkel and Bull, 1979). Although the flower may seem quite different, the household geranium is a close relative of this plant.

habitat . . .
ALLUVIAL. rich woods, thickets and coves
•zone—4a

form . . .
UPRIGHT. dense, slender, HAIRY stemmed herb (1–2′)
•perennial—dark thick rhizome

foliage . . .
OPPOSITE. simple, palmately LOBED (5–7), gray-green leaves (4–5″) with round-toothed margin; long-petioled, pubescent both surfaces

flower . . .
SOLITARY. purplish pink (1–1½″), radially symmetrical, white (2–5) clusters; sepals (5), petals (5), WOOLLY at base
•season—spring, early summer (April–June)

fruit . . .
CAPSULE. elongated, greenish brown (¼–⅜″), BEADED capsule with (⅛″) glabrous seeds; splitting upward from base, hairy
•season—maturing summer

WOODLAND FLOWERS

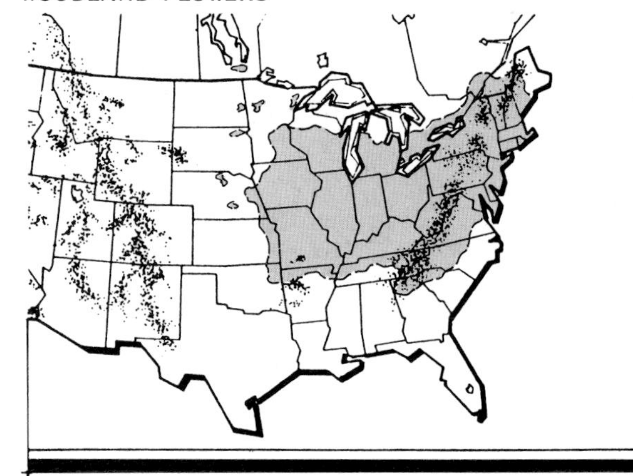

190 *Geranium maculatum* · Spotted Geranium

Sharplobe Hepatica is one of the most beautiful early spring woodland flowers. The blooms resemble corsages. Each flower clump is different in color, ranging from lavender to blue to pink to white. The leaves of Hepatica are liver-shaped. The new leaves, which are shiny green, are often found the following spring flattened to the ground and changed to a red-brown.

It was once thought that a plant resembling an organ of the human body was somehow useful in treating disorders of that organ (Runkel and Bull, 1979). Hepatica has proved to have no medicinal value.

habitat . . .
WOODS. moist, rich woodlands, wooded slopes and mixed deciduous forest
- zone – 4a

form . . .
LOW. dense, stemless, mounded herb (4–6″)
- perennial – short rhizome with above ground stems ABSENT

foliage . . .
ALTERNATE. simple, cordate 3-LOBED, shiny bright green leaves (2–2½″) with entire margin; lobes acute, purplish beneath
- color (fall) PURPLISH bronze
- season – evergreen, PERSISTING through winter

flower . . .
SOLITARY. white-pink or blue (½–1″) on softly elongated (4–6″) HAIRY scapes, surrounded by (3) sessile bracts; petals ABSENT
- season – early spring (March–April) before leaves appear

fruit . . .
ACHENE. pubescent, short-beaked aggregate
- season – maturing late spring

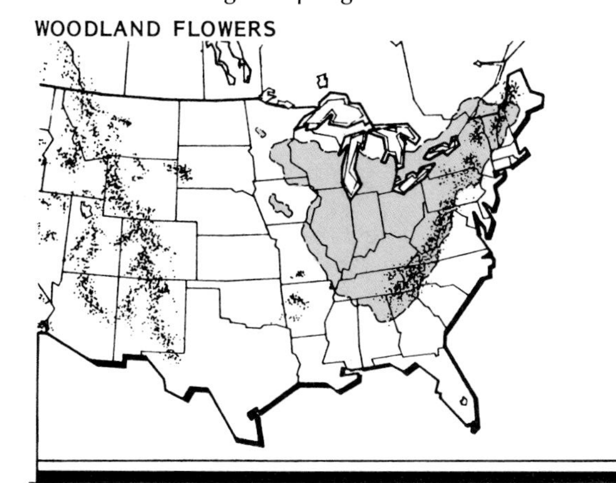

Hepatica acutiloba · Sharplobe Hepatica

Virginia Waterleaf, or John's Cabbage, is found in extensive colonies as a common groundcover in shady woods on rich, moist soils. It blooms from early May to July. Its tight flower clusters are composed of many individual bell-shaped blossoms. Each flower has five stamens that protrude beyond the bell, giving the flower head a hairy appearance. The leaves, divided into five or more segments, are broadly triangular. The name Waterleaf comes from the peculiar light green markings on the leaves, which suggest watermarks on paper. Tender young leaves of Virginia Waterleaf were used as greens by Indians and pioneers (Runkel and Bull, 1979).

habitat . . .
ALLUVIAL. moist, rich wooded slopes, stream banks and clearings
•zone – 3b

form . . .
UPRIGHT. open, erect, many stemmed herb (1–2½′)
•perennial – angled hairy stem with scaly, elongated rhizomes

foliage . . .
ALTERNATE. simple, PINNATELY divided leaflets, (5–7) lanceolate-ovate, bright green leaves (2–5″) with sharp-toothed margin; MOTTLED, water-stained appearance

flower . . .
CLUSTER. white turning violet-purple (¼–½″), BELL-shaped clusters arising in leaf axils on long COILED stalks; HAIRY filaments
•season – spring, early summer (May–Aug.)

fruit . . .
CAPSULE. globular, (⅛″) dia. with tan brown (1–3) seeds
•season – maturing late summer

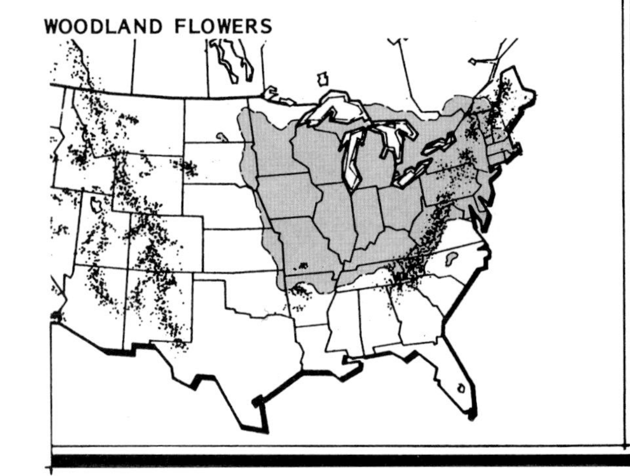

192

Hydrophyllum virginianum · Virginia Waterleaf

Atlantic Isopyrum, or False Rue Anemone, is not an anemone at all, but its flowers are difficult to distinguish from those of Rue Anemone. Both have loose clusters of several white flowers with five petal-like sepals. Atlantic Isopyrum blooms very early, from April to May. It is found in rich, moist bottomland soils, and it grows in large patches in woodlands, swales, and floodplains. Atlantic Isopyrum leaves are divided into three segments and then into three leaflets. Each leaflet is deeply cut, much more so than the leaflets of Rue Anemone.

habitat . . .
WOODS. rich, moist, alluvial limestone woods
•zone – 4a

form . . .
UPRIGHT. open, many-stemmed herb (8–16″)
•perennial – yellowish underground tuber (corms)

foliage . . .
OPPOSITE. ternately COMPOUND (2) leaflets (5–9) ovate, bright green leaves (½–¾″) with entire, 3-LOBED margin

flower . . .
SOLITARY. small, white (½″) clusters with slender stems in leaf axils; axillary, petals ABSENT
•season – early spring (April–May)
•sex – monoecious

fruit . . .
POD. small, reddish; opening along one side
•season – maturing late summer

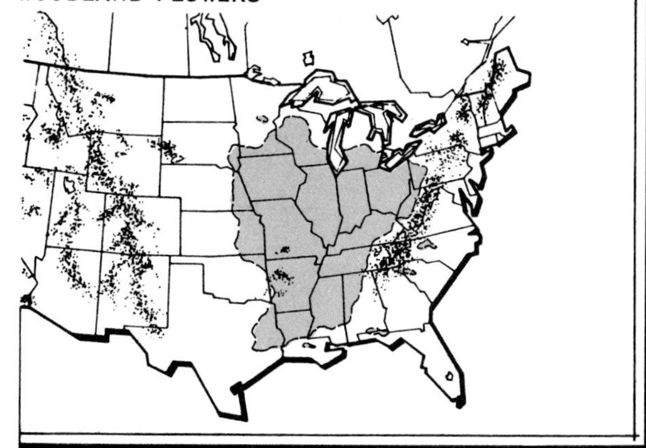

Isopyrum biternatum · Atlantic Isopyrum

Virginia Bluebells, or Virginia Cowslip, is found in rich, moist bottomland soils. It blooms from March to May. The attractive flower is most spectacular when large colonies of Bluebells are seen.

The plant stands erect on fragile stems that support its smooth gray-green foliage. The blossoms appear first as pink buds; then they open into light blue, trumpet-shaped flowers. The four wrinkled nutlets produced from each flower can be planted as soon as they are ripe.

Virginia Bluebells have a short blooming season. By mid-summer, the plant withers and disappears.

habitat . . .
ALLUVIAL. rich, moist woods, rarely meadows
•zone—4b

form . . .
UPRIGHT. ascending, erect, stemmed herb (8–24″)
•perennial—tender, succulent stem and rootstock

foliage . . .
ALTERNATE. simple, oblong, gray-green leaves (2–8″) with entire margin; long-petioled, upper sessile with rounded apex

flower . . .
CLUSTER. long, blue, TRUMPET-shaped (¾–1″), nodding clusters of pink buds with corolla 5-LOBED; terminal
•season—spring, early summer (March–June)

fruit . . .
NUT. wrinkled (4) erect nutlets
•season—maturing early summer

WOODLAND FLOWERS

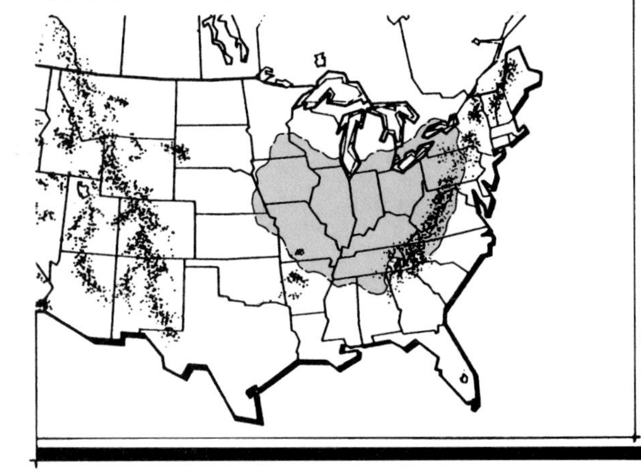

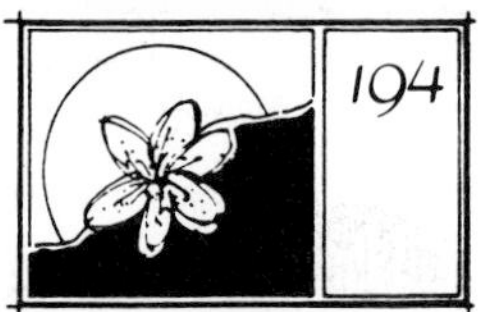

Mertensia virginica · Virginia Bluebells

Interrupted Fern is a rugged and persistent plant; it grows in almost any type of soil and any location, though it seems to prefer dry, stony sites. It is one of the earliest ferns to appear in spring. Like many other woodland perennials, the leaves wither early after the first frost, leaving only its brownish stalk.

The leaves appear to arch as they grow and have distinct interruptions in the center of the leaf stem, hence the name. These interruptions bear the plant's fertile leaflets.

habitat . . .
WOODS. rich woodlands, dry woods edge and rocky shaded slopes
•zone—3b

form . . .
ERECT. spirally arranged, arching coarse-stemmed fern (2–4′)
•rhizome—stout, creeping, without scales; stubby
•leafstock—round, green, stout clusters with semi-grooved face; fertile stalks, longer, more erect
•fiddleheads—stout, woolly, brown; among first to appear in spring

foliage . . .
FROND. pinnately cleft COMPOUND, leaflets (15–20) narrow lance-shaped, bright green leaves (3–6″) sub-alternate with deeply cut oval, blunt-tipped, OVERLAPPING lobes; veins forked, reaching margin
•fertile—fronds interrupted at center by (4–10) pairs bearing spores; dark brown, cylindrical clusters WITHERING rapidly, sterile; fronds curved upward
SPORE. short-lived, green (2–3″) clusters enclosed by dark brown, short-stalked cases; germinating within a few days
•season—deciduous, perennial
•sex—monoecious

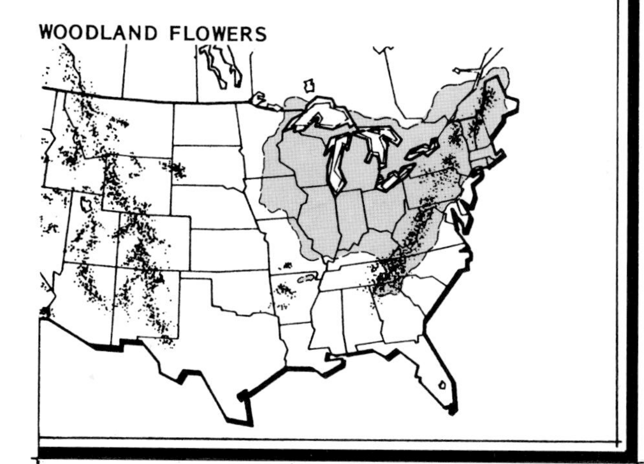

Osmunda claytoniana · Interrupted Fern

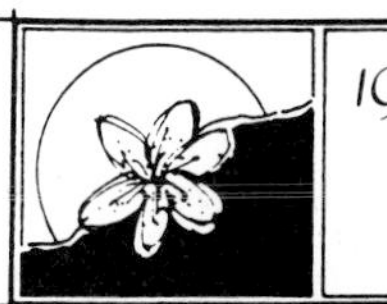

Blue Phlox, or Wild Sweet William, is a woodland flower common to the rich, moist soils of woodland slopes and stream borders. The delicate blue flower is held in loose clusters on sticky stems. It is well known for its color, beauty, and fragrance. It is popular with children because of its timeliness for May baskets, since it blooms from April to June.

Indians used the leaves of some phlox species to brew a tea to treat eczema and to purify the blood. A tea of boiled roots was also used to treat venereal diseases (Runkel and Bull, 1979).

habitat . . .
WOODS. rich, moist woodlands, mixed deciduous forest and open fields, prairies
•zone—4a

form . . .
ERECT. loosely branched, hairy-stemmed herb (12–20″)
•perennial—spreading basal shoots; rhizomatous

foliage . . .
OPPOSITE. simple, ovate-lanceolate, green leaves (1–2″) with entire margin; unstalked

flower . . .
CLUSTER. blue to PINKISH blue (¾–1″), TRUMPET-shaped, in loose clusters; petals (5) united on short branches
•season—early summer (April–June), FRAGRANT

fruit . . .
CAPSULE. 3-valved (⅛–¼″), PAPERY with elliptic one seed
•season—maturing summer

WOODLAND FLOWERS

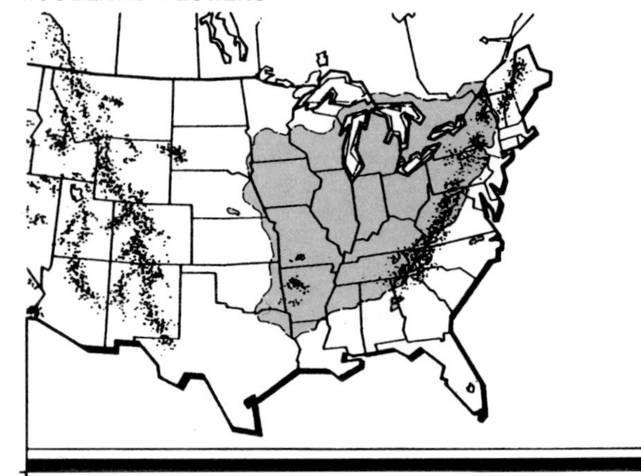

Phlox divaricata · Blue Phlox

Common Mayapple is usually found in colonies in moist soils of open woodlands and floodplain terraces.

The plant's stalks support a single umbrella-shaped leaf; on those where a branch occurs and two leaves are produced, a flower will grow. This white flower blooms in May, on a short stem at the junction of the two leaves. It later produces a green, apple-like fruit.

Common Mayapple's foliage, rootstocks, and green fruit are all somewhat poisonous (Runkel and Bull, 1979). When the fruit is ripe, it is greenish-yellow; it then becomes edible and is used to make preserves.

habitat . . .
WOODS. moist, rich woodlands, mixed deciduous forest and meadows
•zone – 3b

form . . .
ERECT. upright, open-stemmed herb (1–1½′)
•perennial – rhizomatous

foliage . . .
OPPOSITE. peltate palmately LOBED, bright green leaves (8–12″) with toothed margin; segments (5–9) rarely uniform; long-petioled (1–2) leaves

flower . . .
SOLITARY. yellow-white, (2″) dia. in axis of two-leaf petioles; NODDING
•season – early spring (March–April)
•sex – monoecious

fruit . . .
BERRY. small, yellow-green, (⅜–1″) dia., one-celled
•season – maturing late summer, early autumn

WOODLAND FLOWERS

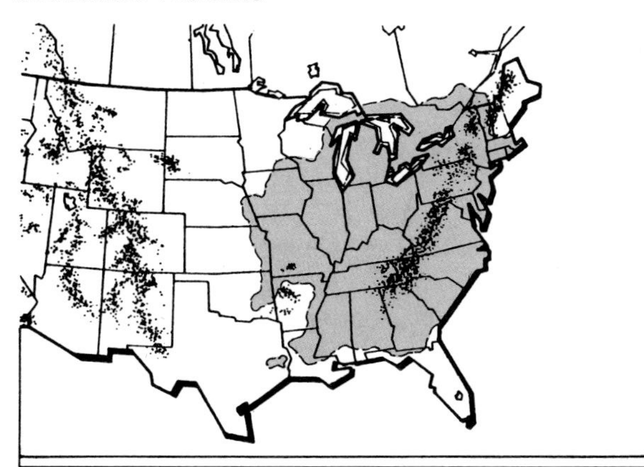

Podophyllum peltatum · Common Mayapple

Common Polypodium, or Wall Fern, grows where rocks and cliffs provide rich, often very shallow, sub-acid soil. It grows luxuriantly with mosses in cool, damp, moist shade along watercourses.

Like most ferns, Wall Fern does not produce typical flowers and seeds for reproduction. Instead, it develops spores on the underside of its leaves that are transported by the wind.

habitat . . .
WOODS. moist, rich woodlands, stream banks and rocky limestone cliffs
- zone – 3a (tropical and semi-tropical)

form . . .
MAT. low, mound-like, spreading, arching stemmed fern (4–10″)
- rhizome – horizontal, widely creeping, thick, densely covered with CINNAMON-colored scales; scarred from KNOB-jointed, withered stems
- leafstalk – smooth, slender green knob at base; 1/3 leaf length
- roots – dark, short, fine, spreading; shallow-rooted
- fiddleheads – red brown, woolly, uncoiling from base

foliage . . .
FROND. pinnatifid COMPOUND leaflets (10–20) oblong lance-shaped, deep green leaves (4–12″) with entire to wavy margin; leathery, once-cut blunt tipped; veins free and forked, not reaching the margin
- season – evergreen

SPORES. round, red-brown fruitdots in two rows, numerous on upper leaflets; NO indusium (growth covering or surrounding sorus or fruitdots)
- season – maturing summer

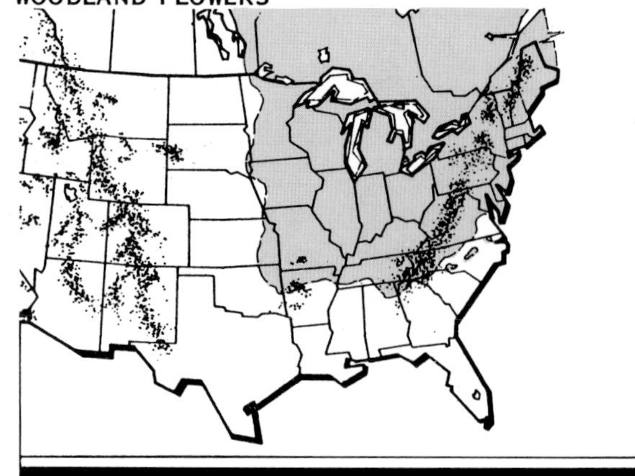

Polypodium vulgare · Common Polypodium

Bloodroot is found in rich, moist, but well-drained soils. It grows 6 to 14 inches tall from a horizontal rootstock. The thick root oozes a bright red juice when cut or broken. Though poisonous, the juice was used by Indians as a dye for fabrics, tools, and war paint (Runkel and Bull, 1979).

As the flower stalk emerges in April or May, it is covered by a single leaf rolled loosely around it. The daisy-like flower is borne on its own stalk, which is initially taller than the leaves. After the flower has bloomed, the leaves unroll, flatten out, and soon grow enough to cover the seed capsule produced by the flower.

habitat . . .
WOODS. mixed deciduous forest, rich woodlands, streams and wooded slopes
•zone—3b

form . . .
ERECT. low, glabrous, upright STEMLESS herb (8–10″)
•perennial—thick rhizome rootstock with acrid, reddish orange, MILKY juice (poisonous)

foliage . . .
SINGULAR. simple, palmately (5–9) LOBED, cordate bluish green leaves (4–7″), lobes rounded with entire margin; petioles often (8″) with basal leaf CURLING around flower stalk

flower . . .
SOLITARY. white (1½–2″), PERFECT with (8–12) elliptic petals on (8″) scapes having numerous golden stamens; embraced by leaf
•season—late spring, early summer (March–May)

fruit . . .
CAPSULE. oblong, green-brown (1″) 2-valved with pointed ends having numerous smooth seeds; black
•season—maturing early summer

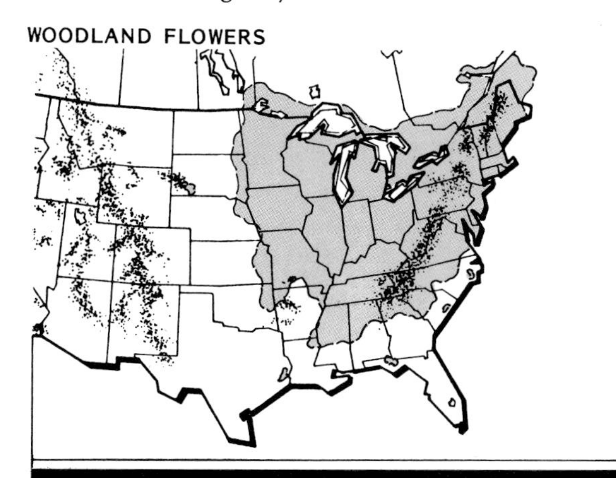

Sanguinaria canadensis · Bloodroot

Feather Solomonplume is found in rich, moist woods, and it blooms from May to July. This plant is often called False Solomon's Seal because its stem and leaves look like those of true Solomon's Seal.

The many small, white flowers in a terminal cluster mature from July to August into numerous brown, speckled berries. The berries turn bright red in the fall. The fruit is edible. Pioneers referred to these berries as scurvy berries, and they probably ate them to treat or prevent that disease (Runkel and Bull, 1979).

habitat . . .
WOODS. rich, moist woodlands, alluvial terraces and ravine, wooded slopes
•zone—3b

form . . .
ARCHING. upright, slender-stemmed herb (1–3′)
•perennial—simple stems from horizontal, knotty rootstock (rhizome); CIRCULAR (yearly) scars

foliage . . .
ALTERNATE. simple, flat-to-spreading lanceolate bright green leaves (2–6″) with entire margin; conspicuously PARALLEL-veined (2 ranked)

flower . . .
CLUSTER. small, white (¼″), dense, showy (3–4″), PYRAMIDAL cluster; terminating
•season—late spring, early summer (May–July)

fruit . . .
BERRY. small, red (¼″), globular cluster; yellowish green at first with brown speckles, later bright red
•season—maturing late summer

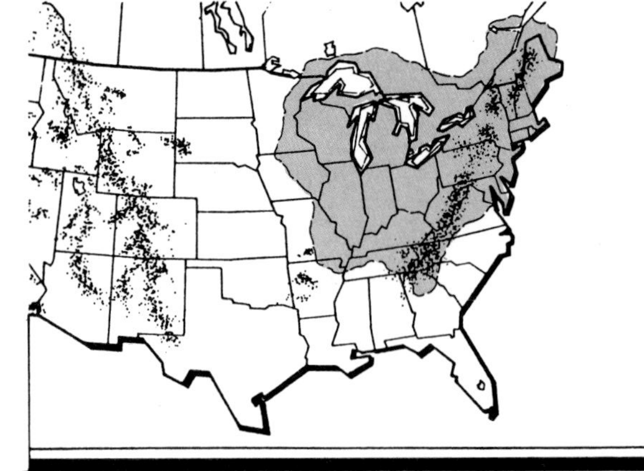

Smilacina racemosa · Feather Solomonplume

Early Meadow Rue, a graceful plant of moist woodlands, has leaves divided into many round, lobed segments similar to those of Columbine. The flowers have golden yellow stamens that hang like small tassels, yet the flowers are without petals and sepals.

The female tassel-like blossoms are purple and appear on separate plants. This accounts for the scientific species name *dioicum,* a Greek word meaning "two households."

habitat . . .
WOODS. rich, moist woodlands, ravines and wooded slopes
- zone – 3b

form . . .
ERECT. slender, glabrous, stemmed herb (1–2½′)
- perennial – smooth, leafy stem with FERN-like leaf appearance

foliage . . .
ALTERNATE. ternately COMPOUND, leaflets (3) orbicular, bluish green leaves (½–2″) with rounded (3–4) lobed entire segments; long-stalked, pale beneath

flower . . .
CLUSTER. (male) yellow (¼–⅜″), open (3–5″), drooping clusters; (female) purplish green, nondrooping
- season – early spring (April–May)
- sex – dioecious

fruit . . .
ACHENE. ovoid, strongly ribbed; one-seeded, sessile
- season – maturing spring

WOODLAND FLOWERS

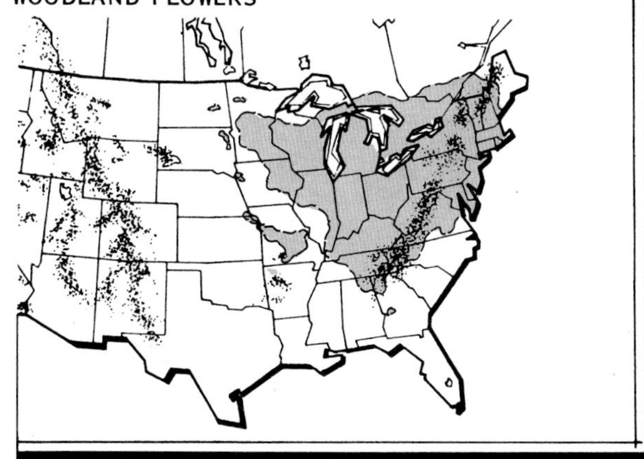

Thalictrum dioicum · Early Meadow Rue

Dwarf Trillium is found in rich, moist woodlands where soils are deep. It is sometimes called Snow Trillium because it can occasionally be found in bloom when there is still snow on the ground. Early pioneers called trilliums Birthroot, because Indians used the plant to induce labor (Runkel and Bull, 1979).

The name trillium comes from the Latin word for "three." The flower parts are all grouped in multiples of three: the white petals, the green sepals, and the leaves. The flower also has six stamens and a three-chambered ovary. The berries that develop usually have three or six angles.

habitat . . .
WOODS. moist, rich woodlands and limestone ledges
•zone – 4a

form . . .
LOW. arching, singular-stemmed herb (2–5″)
•perennial – stem sheathed at base with thick, ascending vertical rhizome

foliage . . .
WHORLED. narrow-ovate (3) gray-bluish green leaves (1–2″) with entire margin; long-petioled, apex tapering to a point

flower . . .
SOLITARY. (3) white (½–1¼″) petals with (3) sepals on short, erect stem
•season – early spring (March–May); one of the earliest spring flowers

fruit . . .
BERRY. globular, 3-ribbed pod
•season – maturing early summer

WOODLAND FLOWERS

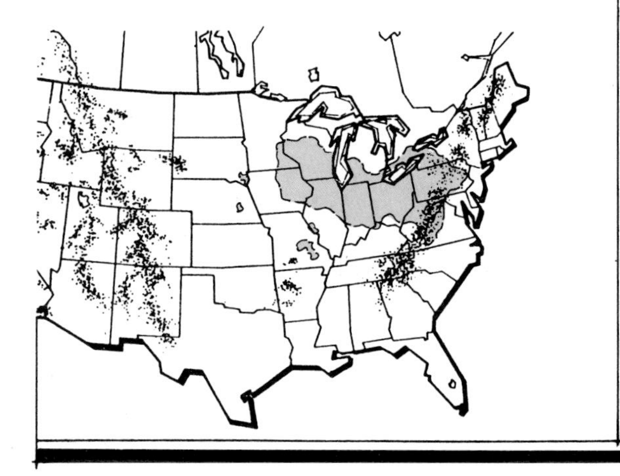

Trillium nivale · Dwarf Trillium

Big Merrybells, or Bellwort, is common to the rich soils of shady upland slopes. When in bloom from April to June, Merrybells can be easily recognized because it is always yellow, drooping, and twisted.

The leaves have no petioles and tend to loop around the stem, scarcely unfolding during flowering. This gives the plant a twisted, droopy appearance.

Merrybells leaves and stems were used by early pioneers for greens. In Europe, the plant was used to treat throat problems since the flower resembles a human uvula hanging from the stalk (Runkel and Bull, 1979).

habitat . . .
WOODS. moist, wooded coves and thickets
•zone – 3b

form . . .
ERECT. simple, singular-stemmed herb (6–18″)
•perennial – short, creeping rootstock

foliage . . .
ALTERNATE. perfoliate, elliptic-obovate shiny green leaves (3–5½″) with entire margin; pubescent beneath

flower . . .
TUBULAR. yellow (1–2″), nodding, BELL-shaped, borne in leaf axils; inner surface smooth; perfect
•season – spring, early summer (April–June)
•sex – monoecious

fruit . . .
CAPSULE. 3-angled with brown, plump (⅛″) seeds (similar to BEECHNUT)
•season – maturing in late spring

WOODLAND FLOWERS

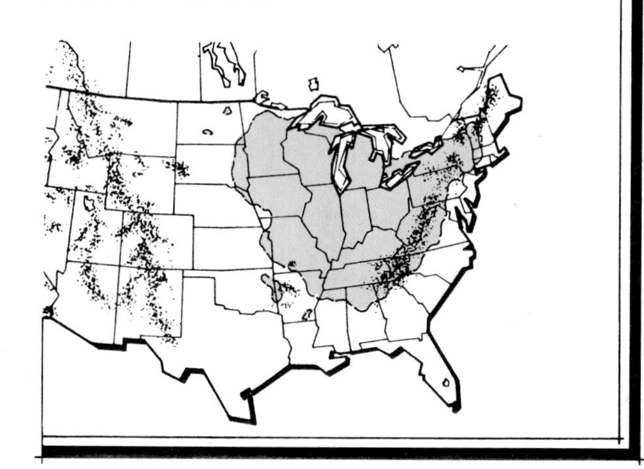

Uvularia grandiflora · Big Merrybells

Violet species abound throughout the Midwest. They frequent a wide variety of habitats, from wet to dry soils and from woodlands to prairies.

Violets typically bloom from April to June and usually grow no taller than 3 to 5 inches. The leaves vary in shape, but are usually heart-shaped.

Identification becomes highly technical, since violets frequently hybridize to produce blooms ranging from blue to purple to white to yellow. They are probably the most common wildflower of the woodlands; and quite possibly, they are the most loved.

habitat . . .
WOODS. moist, rich woodlands, mixed deciduous forest, open meadows and stream banks
•zone – 3b

form . . .
LOW. upright to mounding, leafy stemmed herb (3–8″)
•perennial – short, dark, scaly horizontal underground rootstock (rhizome)

foliage . . .
ALTERNATE. simple, cordate, bright green leaves (2–5″) with rounded, scalloped margin; long petioled, prominently PALMATELY veined

flower . . .
SOLITARY. 5-lobed blue, purple, yellow to white (½–¾″), nodding on slender scape; may have veins on lateral petal, lower petal spurred
•season – late spring, early summer (March–June)

fruit . . .
CAPSULE. boat-shaped, 3-valve, splitting upward; seeds obovate
•season – maturing summer

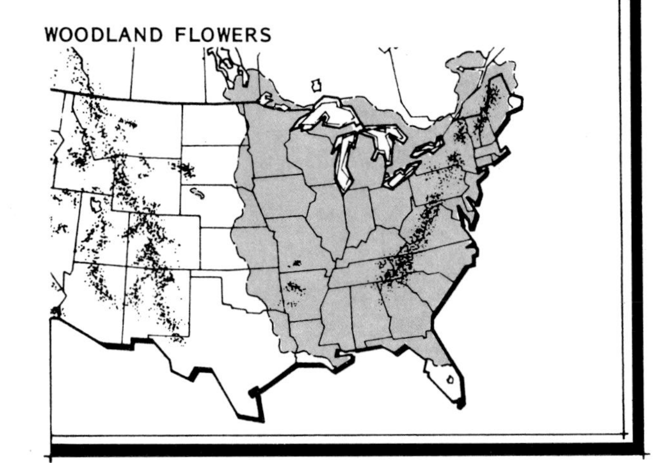

Viola species · Violets

The Pine-Fir-Birch community of the Upper Midwest is a remnant community from a bygone era. Today this community is largely found in Canada and the Great Lakes region and is referred to as the Boreal Forest in most plant association studies.

Until the Wisconsin glacier receded between 12,000 and 15,000 B.C., the upper midwestern region had shown the effects of millions of years of weathering and erosion. Only a few sites escaped the leveling influence of this glaciation. Narrow gorges, valleys with steep water-worn cliffs, towering picturesque rock pinnacles, caves, and springs are typical of the terrain where the remnants of this community are found today.

Mixed or pure stands of White Pine and Paper Birch are frequently found on shallow soils over limestone bedrock. These steep, moist, north-facing slopes are constantly chilled by cold air drainage and cold water seepage through caverns in the limestone. A few rare stands of Balsam Fir still exist, with a thick mat of entangled Canada Yew sometimes covering whole slopes.

Pine-Fir-Birch Community

Trees: Canopy (over 48′)

Shrubs: (less than 6′)

PINE-FIR-BIRCH COMMUNITY

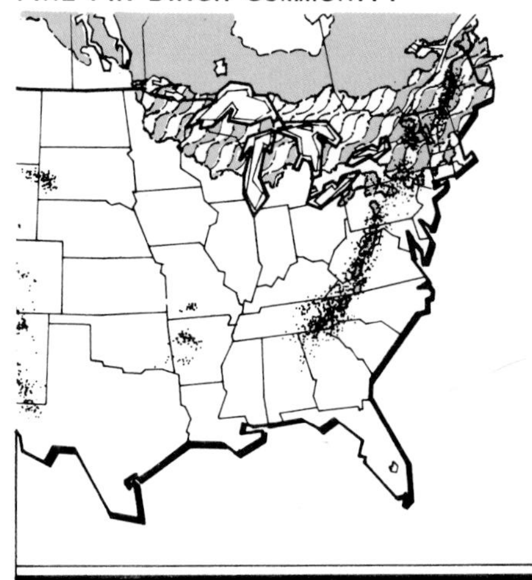

Balsam Fir is essentially a tree of the great northern forests and may be its most popular tree. A delicious fragrance is released by the tree's needles after they dry; for this reason, boughs of Balsam Fir have long been used to decorate homes at Christmas time. The foliage is also used in the manufacture of so-called pine pillows, which are cloth bags filled with the fragrant needles (Grimm, 1983).

Balsam Fir can be found on mountain slopes where there is cold air drainage or in cold swamps and bogs. It is a hardy tree in its native woods and a handsome, symmetrical tree in its youth, but it is short-lived and loses its lower branches as it matures. The needles are short, growing spirally on the twig and yet standing upright. Like all firs, the cones of the Balsam Fir are held upright and disintegrate on the tree.

Chickadees and Purple Finches are among the many birds that feed on Balsam Fir seeds. Those not eaten fall with the cone scales to the ground below. Deer frequently browse on the foliage, as they prefer balsam thicket swamps for their winter haunts (Elias, 1980).

habitat . . .
FOREST. lowland wet-mesic, lowland wet, upland mesic, stream, water edge, bogs, north and east facing slopes and moist, rocky lands
- zone–2

form . . .
CONICAL. narrow, pointed, SPIRE-like, small canopy tree (50–75′)
- branching–stiff, horizontal limbs often spreading along ground
- twig–stout, yellow, reddish green, smooth with circular leaf SCAR having clustered, blunt, resinous buds; yellow, red-brown, in GROUPS of 3
- bark–smooth, ash gray-brown, shallow, furrowed with RESINOUS blisters

foliage . . .
NEEDLE. flat, linear, 2-RANKED at right angle, light glossy green leaves (½–1″) with whitish LINES underneath; curving upward, rounded to short, pointed tip
- color (fall) dark GREEN
- season–evergreen, AROMATIC

flower . . .
CONE. (male) yellow, pinkish green, small; (female) pale yellow (¾–1″), erect
- sex–monoecious

fruit . . .
CONE. upright, PURPLE-brown, FLESHY (2–4″) with central stalk of SHEDDING scales on uppermost twigs
- season–maturing late summer, falling apart upon ripening

PINE-FIR-BIRCH COMMUNITY

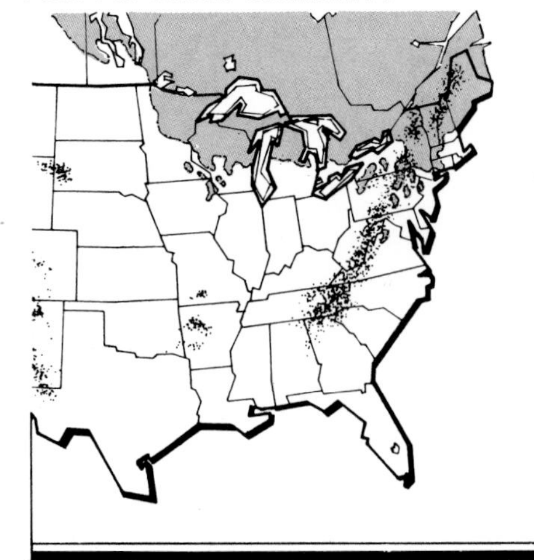

Abies balsamea · Balsam Fir

Yellow Birch is a tree of the northern woods that grows taller than any tree in Canada today except White Pine (Borland, 1983). It is a tree that demands coolness. In the mid-Atlantic states, it grows on wet, swampy ground. In the Appalachian and New England hills, it is abundant around ravines where its visible roots lay hold of the rocky hillsides like monstrous yellow claws.

The bark, like that of most birches, is the main point of interest. It has gold and silver curls that are so thin and translucent that they glitter even in shadows. The tree is at its best when highlighted by sunlight or snow. Strips of the bark peel off easily and will burn even when wet. In fact, Yellow Birch bark can be used to start a fire in a downpour.

The leaves and catkins are very much like those of other birches, and the fall color is the bright, clear yellow common to the family. The wood is valuable, and it is used today for interior finishes in many new homes.

habitat . . .
FOREST. upland mesic, lowland mesic-wet, north and east facing slopes
- zone–3a

form . . .
OVOID. conical, small canopy tree (50–75′)
- branching–stout, ascending limbs
- twig–smooth, yellow-brown, slender, wintergreen aroma with stalked, sharp-pointed, chestnut brown buds; terminal bud absent
- bark–thin, yellow, orange-BRONZE, exfoliating, papery CURLS with glossy, horizontal lenticels

foliage . . .
ALTERNATE. simple, oblong-ovate, dark green leaves (1¼–2″) with DOUBLE-toothed margin; pale, yellow-green beneath with tufts of hair at vein axils
- color (fall) YELLOW-gold
- season–deciduous

flower . . .
CATKIN. (male) pendulous, yellow-green (2–3″), PENCIL-like clusters; (female) greenish, ERECT (½–¾″) in back of leaves on same twig
- sex–monoecious, male preceding autumn

fruit . . .
STROBILE. green becoming tan-brown, ERECT (1–1½″), unstalked; PAPERY
- season–appearing early summer, maturing autumn

PINE-FIR-BIRCH COMMUNITY

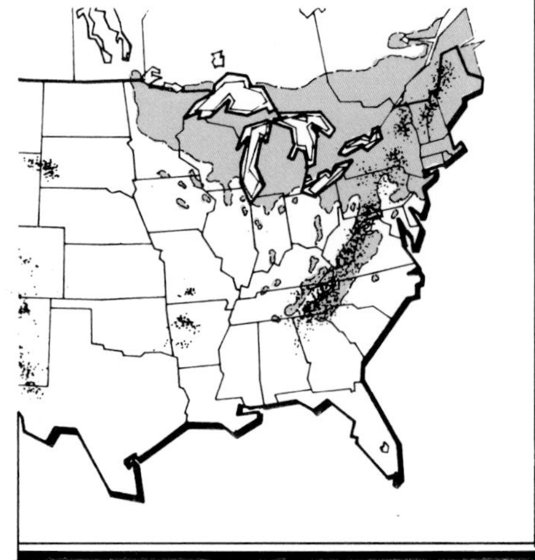

Betula lutea · Yellow Birch

Paper Birch is primarily an understory tree of the northern woods, usually found on cool, north- and east-facing, rocky slopes. It is a tree so unlike any other that everyone knows it: Paper Birch is the tree pictured in almost every beer commercial or mountain scene.

Indians used Paper Birch bark to make their lightweight canoes. The bark, peeled in great sheets from 2- to 3-foot trunks, was shaped to frames of cedar or spruce and bound in place with rootlets. The seams were then smeared with pitch to seal the joints (Borland, 1983).

The bark, although soft and smooth to the touch, is one of the most durable plant substances. One birch log was buried in Siberia for so long that the wood had turned to stone, but the fossil log still wore its birch bark unchanged through the centuries (Platt, 1968). Like the bark of Yellow Birch, Paper Birch bark is resinous and so flammable that there is no better material with which to kindle a campfire. The bark, when removed from the tree, reveals a red inner bark that later turns black. Many states have outlawed removing bark from birches because it never grows back.

Paper Birch is a beautiful accent tree when used with a background of evergreens. It grows at a medium to fast rate, is short-lived, and does not do well south of the northern tier of states.

habitat . . .
FOREST. upland mesic and mesic-dry, steep rocky land, north and east facing slopes
- zone—2

form . . .
IRREGULAR. globular, small canopy tree (50–75′)
- branching—horizontal, ASCENDING limbs
- twig—smooth, reddish brown, glossy, slender WHITE lenticels with tan-brown buds; stalked
- bark—creamy, chalky WHITE, exfoliating horizontally in PAPERY strips; orange inner bark, furrowed, black scaly base

foliage . . .
ALTERNATE. simple, oblong-ovate, thin, dark green leaves (2–4″) with SHARP, double-toothed margin; smooth, whitish beneath with 5–9 veins on each side
- color (fall) lemon-YELLOW
- season—deciduous

flower . . .
CATKIN. (male) PENDULOUS, yellow-green (3–4″) drooping clusters; (female) greenish, ERECT (1–1½″), appearing with leaves
- sex—monoecious, male PRECEDING autumn

fruit . . .
STROBILE. green becoming tan-brown, cylindrical (1–1½″) clusters; pendulous
- season—appearing summer, maturing autumn

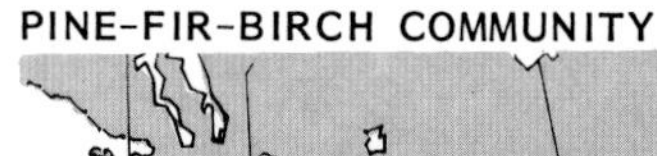

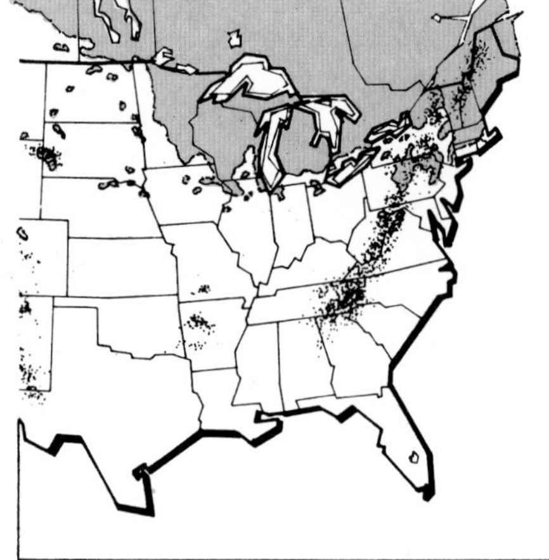

Betula papyrifera · Paper Birch

Eastern White Pine is a very popular evergreen, due primarily to its soft, plume-like needles, which are grouped in bundles of five, and its horizontal branching form. It is the tallest of the northeastern conifers, growing to 100 feet and sometimes much taller.

Years ago, White Pine was an outstanding source of fine lumber in the North, but harvesting has been so thorough that a single, mature, perfect specimen is difficult to find today.

White Pine is among the finest evergreens for ornamental planting. It grows rapidly when young—a seedling can reach a height of more than 50 feet in 30 years—and it accepts severe pruning better than any other tree of its general type. This last characteristic makes it useful in tall hedges and windbreak plantings. White Pine is slow to bear cones but does so after its twentieth birthday and fits the old adage of "free, white and twenty-one" (Leopold, 1977).

Eastern White Pine prefers moist sandy loams but will grow on a variety of sites and soil conditions. Among its many uses, it is often cultivated for Christmas trees.

habitat . . .
FOREST. upland mesic, steep rocky land and north facing slopes
- zone—3b

form . .
OVOID. conical, large canopy tree (75–100′)
- branching—horizontal, TIERED limb pattern
- twig—SLENDER, green, gray-brown with ovoid, sharp-pointed buds; RESINOUS
- bark—smooth, gray-brown, deep furrows of RECTANGULAR plates

foliage . . .
NEEDLE. slender, soft, glossy dark green leaves (2½–5″) in groups of 5s; AROMATIC
- color (fall) bright GREEN
- season—evergreen

flower . . .
CONE. (male) small, YELLOW-green, dense clusters; (female) PINKISH green (3/8″), SOLITARY to few in cluster
- sex—monoecious

fruit . . .
CONE. tan-brown, cylindrical (4–8′), slightly CURVED with rounded RESINOUS scales; pendulous
- season—appearing summer, maturing second year

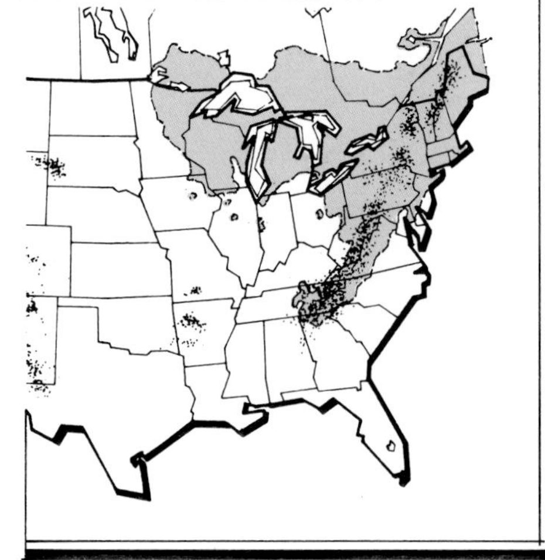

Pinus strobus · Eastern White Pine

Dwarf Bushhoneysuckle is a low shrub or groundcover common to the moist woodlands in northeastern states and as far west as Wisconsin and south into Georgia. It is a plant of dry, shady, and usually rocky hillsides. It resembles the honeysuckle that has opposite leaves and tubular flowers held in clusters. True honeysuckles, however, have leaves that are not toothed and fleshy fruit, whereas Dwarf Bushhoneysuckle has toothed leaves and a beaked fruit capsule. This dry capsule is usually bitter but is very attractive to birds and other wildlife. The yellow, tubular flowers seem odorless to humans, but bees swarm to them when they open in late spring. Both the roots and leaves of this plant were used by Indians to treat everything from sore throats to gonorrhea (Runkel and Bull, 1979).

habitat . . .
FORESТ, SAVANNA. upland mesic and dry, north and east facing slopes
- zone—3a

form . . .
GLOBULAR. open, small shrub (1–3′)
- branching—upright with rounded, loosely suckering limbs
- twig—slender, gray-brown, with long-pointed, 4-SCALED buds; leaf scars with ridges running downward
- bark—tan-brown, slightly exfoliating, PAPERY ridges; inner bark orange

foliage . . .
OPPOSITE. simple, ovate-lanceolate, green leaves (2–5″), leaf base WEDGE-shaped with fine-toothed margin; leaf scars connecting
- color (fall) greenish YELLOW
- season—deciduous

flower . . .
TUBULAR. yellow, FUNNEL-shaped, (3/8–1/2″) dia., clustered in leaf axils
- sex—monoecious

fruit . . .
CAPSULE. tan-brown, long-beaked (5/8″), 2-VALVED, terminal clusters; seeds numerous, small, angular
- season—maturing late summer

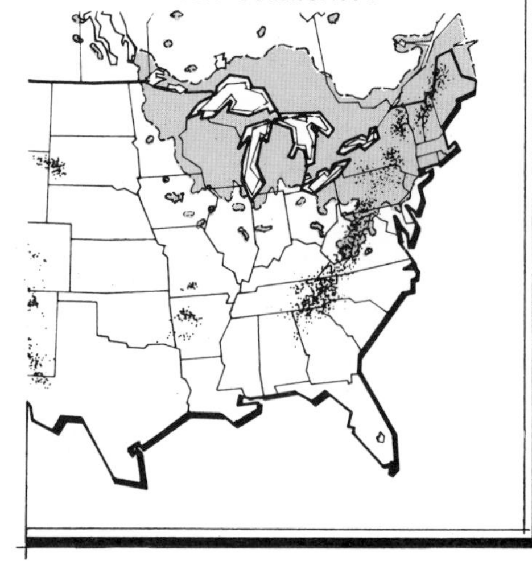

Dierville
lonicera · Dwarf Bushhoneysuckle

Common Juniper, a northern evergreen common to Cedar Glades of dry, barren, stony soils, is a low-spreading shrub, 3 to 6 feet tall. It often forms circular mats several yards in diameter.

The leaves, whorled in sets of three, are awl-like, concave needles having a white stripe down the middle. The needles form sharp points.

The fruit is berry-like, blue-black, and covered with a white, powdery bloom. Sweet and aromatic, it usually contains three seeds. The berries are a major food source for wildlife and have been used medicinally and for the making of gin.

Common Juniper is a long-lived plant that is not tolerant of shade or flooding but prefers sunny, dry exposures.

habitat . . .
FOREST, SAVANNA. lowland mesic-wet, steep rocky land and woods edge
- zone–2

form . . .
IRREGULAR. dense, spreading shrub (3–6′)
- branching–horizontal with low PROSTRATE limbs
- twig–slender, YELLOW-green, smooth, 3-angled
- bark–thin, REDDISH brown to gray, scaly to exfoliating, PLATEY stripes

foliage . . .
NEEDLE. stiff, sharp-pointed, AWL-shaped, 3-sided, blue-green leaves (3/8–1/2″) set in whorls of 3s spreading at right angles; WHITISH band upper, yellow-gray beneath
- color (fall) bluish GREEN
- season–evergreen

flower . . .
CONE. (male) YELLOWISH green, small (1/8–3/8″), ovoid; (female) green (1/4″), globular along leaf axils
- sex–dioecious

fruit . . .
BERRY. globular, BLUISH black, (1/4–3/8″) dia., short-stalked with WHITISH bloom; AROMATIC
- season–maturing second and third year, remaining attached

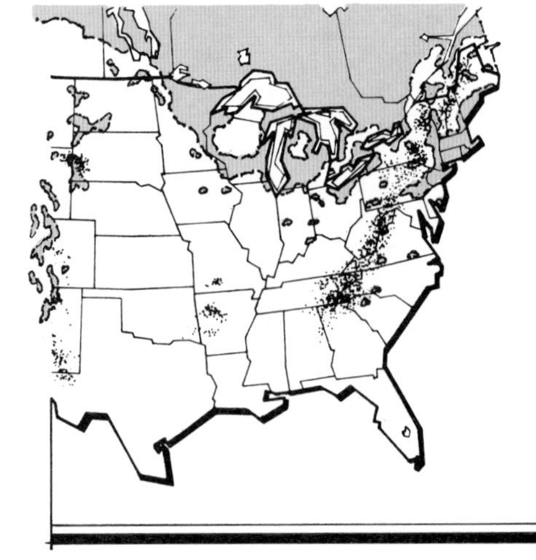

habitat . . .
FOREST. upland mesic, steep moist ravines, limestone cliffs and cool north slopes
- zone – 2

form . . .
PROSTRATE. low mounding or straggling shrub (3–6′)
- branching – stout, spreading limbs with ASCENDING tips
- twig – smooth, green, (old) turning red-brown with small blunt buds; yellow-brown
- bark – thin, reddish brown, scaly to EXFOLIATING in irregular plates

foliage . . .
NEEDLES. linear, flattened, 2-ranked, glossy dark green leaves (¾–1″) with yellowish green bands (2) beneath; sharp-tipped, flexible
- color (fall) GREEN with reddish tint
- season – evergreen, POISONOUS, aromatic

flower . . .
CONE. (male) yellow, small (¼″), globular clusters; (female) tiny, yellow-green, inconspicuous in upper leaf axils
- sex – dioecious

fruit . . .
BERRY. small, SCARLET-red (¼–⅜″), translucent, FLESHY cup-shaped disc with small, blunt-pointed, brown seed
- season – maturing autumn, POISONOUS

PINE-FIR-BIRCH COMMUNITY

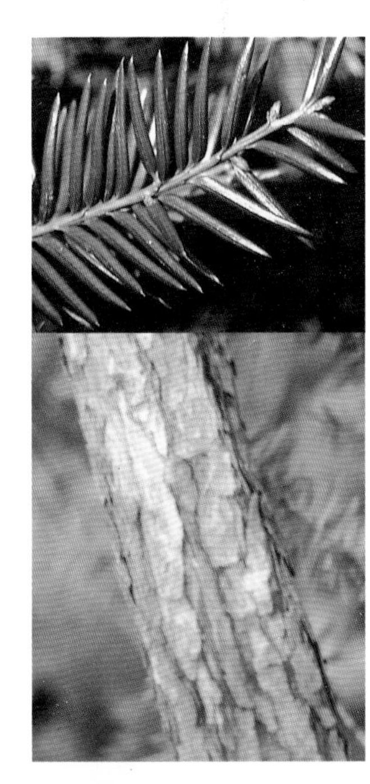

Canada Yew is a low shrub or groundcover found in the deep shade on cool, moist, sloping land. It often grows in tangled mats mixed with ferns and mosses in isolated communities found as far south as northeastern Iowa.

Canada Yew is an evergreen with flat, narrow needles that are dark green above and pale green on the lower surface and are said to be poisonous.

The most attractive feature of the female plant is the bright red, berry-like fruit that grows at the tips of the branches. The fruit, appearing in late summer, has a dark brown, bony seed that is almost completely surrounded by a red, pulpy protection. Many birds are attracted to the fruit and feed on the sweet, reddish pulp.

Taxus canadensis · Canada Yew

American Cranberrybush Viburnum is one of the most handsome plants for the shrub border. It grows from 4 to 10 feet tall on low ground bordering cool streams or swamps. It also grows equally well on drier, more open sites.

It is a favorite shrub for landscape use because of its year-round beauty. American Cranberrybush has large flat-topped flowers that bloom in May or June. These flower clusters have an outer ring of showy, sterile flowers. The small, perfect, fruit-bearing flowers form the center of the cluster. The leaves are three-lobed and maple-shaped, and they turn a deep red each fall.

The drooping clusters of brilliant red fruits of Cranberrybush Viburnum often remain on the branches long after the leaves have fallen. They are tart and acidic and are usually avoided as food by birds until late winter when the berries have lost some of their acidity and other food is scarce. Then Cedar Waxwings and other birds find them appealing, eating most of the berries in just a few days.

Few plants offer as much color and interest throughout the seasons as does American Cranberrybush Viburnum. It is, therefore, highly recommended for use in any design where emphasis and variety are needed.

habitat . . .
FOREST. upland mesic, cool ravines, north and east facing slopes
- zone—2

form . . .
OBOVATE. globular, large shrub (6–15′)
- branching—upright and slightly ARCHING limbs
- twig—hairless, gray-brown with glabrous, RED-brown buds; 2-scaled
- bark—dark gray, slightly fissured to CORKY appearance

foliage . . .
OPPOSITE. simple, ovate, 3-LOBED, bright green leaves (2–5″) with long-POINTED entire lobes; glabrous beneath, grooved petiole
- color (fall) red-MAROON
- season—deciduous

flower . . .
CLUSTER. small, WHITE, flat-topped, (2 to 3½″) dia. with large sterile cluster surrounded by FERTILE outer border; terminal
- sex—monoecious

fruit . . .
BERRY. globular, scarlet-RED, (¼–⅜″) dia. in drooping clusters
- season—appearing summer, maturing fall and PERSISTING through winter (edible)

PINE-FIR-BIRCH COMMUNITY

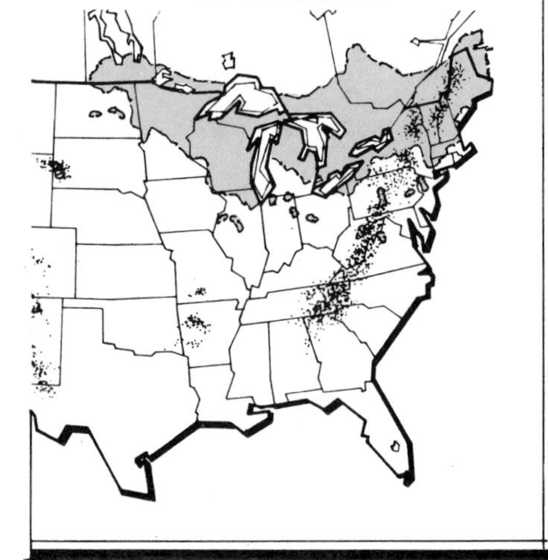

Viburnum trilobum
American Cranberrybush Viburnum

Mixed Floodplain Community

Maple, hackberry, elm, and ash are all found in the Mixed Floodplain community. This community occupies the higher elevations along stream and river corridors; therefore flooding is not as frequent nor as prolonged as it is for the River–Lake Margin community.

In the 1800s, many of America's street trees were selected from the floodplain community. Trees were probably selected for their ability to grow quickly and for their tolerance of adverse conditions, such as flooding and drought.

The Mixed Floodplain community is characterized by a scattering of giant trees that provide a vaulted ceiling, often 100 or more feet tall. Shrubs and understory trees are not common because of flooding, but the early spring woodland flowers are not discouraged, especially at the higher elevations.

Vines do well in this community. Using large trees for support they seem to cling to every tree, giving the Mixed Floodplain community the appearance of a jungle.

Trees: Canopy (over 48′)

Trees: Understory (12–48′)

Shrubs: (6–12′)

Vines:

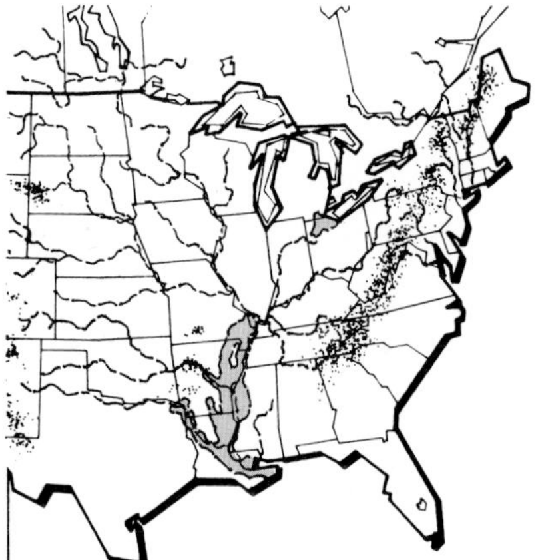

Bitternut Hickory is the most widely distributed hickory throughout the Midwest. It holds its golden leaves longer in the autumn than the other hickories do. This fact, along with its graceful, drooping branchlets, makes it most desirable as an ornamental tree.

Bitternut Hickory also has a great capacity to adapt to varied conditions. It is usually found in moist soils along rivers on swampy ground and is sometimes referred to as Swamp Hickory. It can also be found growing beautifully on dry gravelly ridges.

Bitternut Hickory has pinnately compound leaves, true of all hickories, but it possesses several other characteristics that are peculiar to this tree. The buds in winter are golden yellow, and the nuts have a thin, four-ridged husk, while the kernels are so bitter even squirrels won't eat them. Also, the bark is a rather smooth gray and is interwoven in an attractive pattern. The wood is not as strong as other hickory wood, but it is good for smoking hams.

Bitternut Hickory is a large upright tree with a straight trunk. It grows 75 to 100 feet tall and is very hardy, but it will not live as long as other hickories do. Its maximum age is only about 200 years. Because of its graceful appearance and golden leaves in autumn, Bitternut Hickory should be given high priority as an ornamental shade tree for home plantings.

Carya cordiformis · Bitternut Hickory

habitat . . .
FOREST. upland mesic, upland mesic-dry, lowland wet-mesic, moist-dry slopes
- zone – 4a

form . . .
GLOBULAR. large canopy tree (75–100′)
- branching – upright, ascending limbs with straight trunk
- twig – stout, gray-brown, slender with long-pointed pubescent bud; SULFUR yellow
- bark – GRAY, smooth to shallow-furrowed, scaly ridges; limbs often black

foliage . . .
ALTERNATE. simple, pinnately COMPOUND leaflets (7–11), lanceolate-ovate, YELLOW-GREEN (3–6″) with fine-toothed margin; stalkless, slightly hairy beneath
- color (fall) YELLOW-brown
- season – deciduous

flower . . .
CATKIN. cylindrical, yellow-GREEN (2–3″), drooping clusters in groups of 3s; pendulous
- sex – monoecious

fruit . . .
NUT. round, yellow-green, (¾–1¼″) dia. enclosed by a thin, 4-WINGED, pubescent HUSK (non-edible)
- season – maturing late autumn

MIXED FLOODPLAIN COMMUNITY

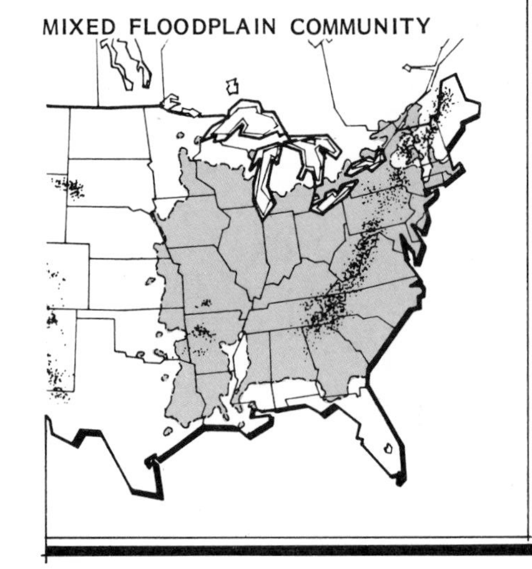

Common Hackberry is a large canopy tree, 75 to 100 feet tall, that is often confused with American Elm. Elm and hackberry are in the same family, and both were gathered from floodplains to be used as street trees at the turn of the century. The fact that some cities have large hackberries as street trees is probably because they thought they were planting elm saplings but got hackberries instead. This mistake was advantageous, since elms have largely disappeared because of Dutch Elm disease.

Common Hackberry is not difficult to identify. It does not have the majestic vase-like appearance that made the elms so popular. Common Hackberry is less predictable in appearance and more erratic in growth, but it does maintain a long, slender trunk and a rounded mass of foliage.

The easiest way to differentiate is to look closely for fruit and to check the bark. The fruit is a small, round berry with a thin, sweet skin and a hard center. Hackberry fruit is nearly always found on the tree, either green in summer or brown in winter. The bark is the tree's most noticeable feature. It is thick and rough with knobby, cork-like ridges. Very rarely does one find a hackberry with even a moderately smooth trunk.

Hackberry grows at a medium to fast rate. It is easily transplanted and matures between 100 and 200 years.

habitat . . .
FOREST, SAVANNA. upland mesic-dry, lowland wet-mesic, floodplains and steep rocky slopes
- zone – 3a

form . . .
GLOBULAR. umbrella-shaped, large canopy tree (75–100′)
- branching – upright, spreading limbs
- twig – gray-brown, slender, ZIG-ZAG with small, egg-shaped buds; light brown, often deformed with bushy growth of witches-broom
- bark – thick, light gray-brown ridges with irregular WART-LIKE knobs

foliage . . .
ALTERNATE. simple, OBLONG-ovate, blue-green leaves (2–5″) with uneven base; sharp-toothed, densely wooly, rough (young)
- color (fall) lemon-YELLOW
- season – deciduous

flower . . .
CLUSTER. small, YELLOW-green, (1/8″) dia. in pendulous clusters
- sex – monoecious

fruit . . .
BERRY. fleshy, PURPLE-brown, globular, (1/4–3/8″) dia. on slender stalk
- season – maturing late autumn

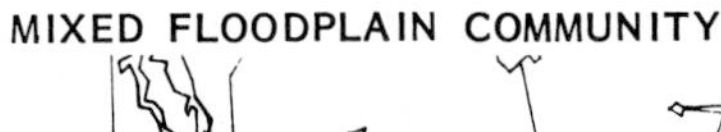

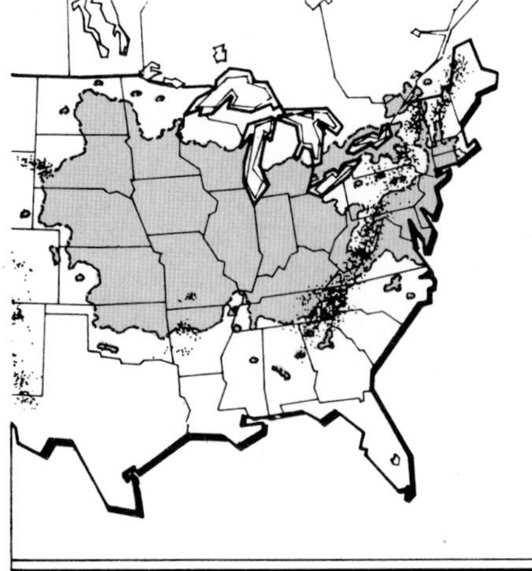

Celtis occidentalis · Common Hackberry

Green Ash has become the most popular tree planted in parks and residential yards throughout the Midwest. It grows at an average rate of 2½ feet a year and reaches maturity as a small canopy tree at 50 to 70 feet. Yet, unlike many other fast-growing trees, it lives from 100 to 150 years.

Green Ash is also adaptable, being very tolerant of moisture and temperature extremes. It prefers moist, deep soils in floodplains but has been used extensively on dry plains in shelterbelts because of its ability to resist drought.

It is easy to see why Green Ash has become so popular. In addition to its aforementioned qualities, Green Ash has a handsome, round-topped silhouette, slender branches, and a beautiful yellow fall leaf color. The deep, dark green of the leaves in summer also provides an attractive, lustrous appearance. The bark, as in other ash varieties, has long, diamond-shaped furrows that help identify the tree when the leaves have fallen in winter.

Green Ash has long been considered a variety of Red Ash. Red Ash in the East is easily identified by its fuzzy twigs on new growth. But in the Midwest, there seems to be a blending of the two, with both Green and Red Ash having varying amounts of fuzz on new twig buds.

The lumber of Green Ash is used for paddles and tool handles as is White Ash, but White Ash is much preferred.

habitat . . .
FOREST, SAVANNA. upland mesic-dry, lowland wet-mesic, stream and lake margin and farmstead planting
- zone–2

form . . .
IRREGULAR. ovoid, small canopy tree (50–75′)
- branching–coarse, recurving limbs
- twig–smooth, rigid, yellow-brown; slender with red-brown, WOOLLY buds set above large leaf scar
- bark–ash gray-brown, shallow-furrowed with DIAMOND-like pattern plates

foliage . . .
OPPOSITE. pinnately COMPOUND, leaflets (7–11) oblong-LANCEOLATE, dark green leaves (3–5″) with fine-toothed margin; glossy upper surface
- color (fall) YELLOW-orange
- season–deciduous

flower . . .
CLUSTER. small, yellowish-GREEN (⅛″), compact clusters
- sex–dioecious

fruit . . .
SAMARA. single-WINGED key, tan-brown (1″), dense, drooping clusters; wing oblong, extending to base of flat body
- season–maturing in late summer

MIXED FLOODPLAIN COMMUNITY

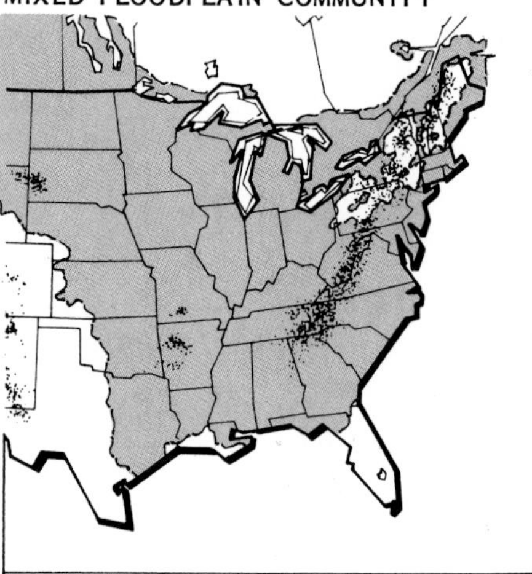

Fraxinus pennsylvanica lanceolata · Green Ash

Blue Ash is a small, round-topped canopy tree that is similar to other ashes in many respects. But it has several distinguishing characteristics that set it apart from other members of the Ash Family. Like Green Ash, it grows 50 to 75 feet tall and is a fast-growing tree. Yet it may live 100 to 150 years and is tolerant of various climates and soils. It grows in the rich bottomlands of fertile floodplains, but some of the finest specimens grow on dry limestone hills in the Great Smoky Mountains.

The fruit and leaves of Blue Ash are similar in shape to Black Ash, but this is where all similarity stops. The most unusual trait of Blue Ash is the four-angled character of its young twigs. These twigs have four distinct, winged, cork-like ridges running lengthwise along the bark; its scientific name, *quadrangulata,* refers to this unusual characteristic. This winged appearance, however, disappears as the twigs become older.

In summer, the inner bark of the tree reveals blue coloring when mixed with water. Many youngsters have been amazed at the beautiful blue that can be achieved just by stirring a crushed twig in clear water. This gives reason for the common name and for the use of the wood in pioneer days for coloring homespun clothing (Werthner, 1935).

Less obvious distinctions are the flowers and bark. The bark has a scaly appearance rather than the diamond-shaped fissures so common to other ashes. Also, Blue Ash is the only ash with perfect flowers, having both stamens and pistil in the same blossom.

habitat . . .
FOREST, SAVANNA. upland dry, lowland wet-mesic, dry rocky slopes, bluffs, ravines and creek edge
- zone – 5a

form . . .
IRREGULAR. narrow, small canopy tree (50–75′)
- branching – coarse, RECURVING limbs
- twig – stout, 4-angled, yellow-brown, having slightly WINGED lines with buds on large leaf scar; tan-brown
- bark – tan-gray with loose, plate-like scales; often EXFOLIATING (old)

foliage . . .
OPPOSITE. pinnately COMPOUND, leaflets (5–11) oblong-LANCEOLATE, yellow-green (2½–4½″) with coarse-toothed margin; long-pointed apex often with tufts of hairs on midrib beneath
- color (fall) YELLOW-gold
- season – deciduous

flower . . .
CLUSTER. small, deep PURPLE (⅛″), compact, open clusters
- sex – monoecious

fruit . . .
SAMARA. tan-brown, single-WINGED key, (1½–2″) dense, drooping clusters; oblong wing extending to base of flattened body
- season – maturing in autumn, PERSISTING through winter

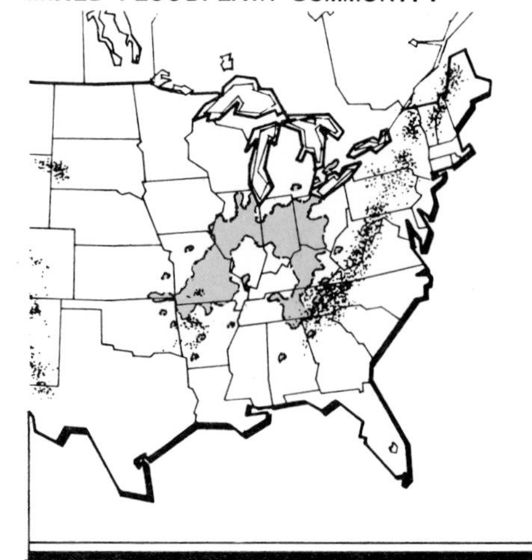

Fraxinus quadrangulata · Blue Ash

Kentucky Coffeetree, when crowded by other trees on rich soil in floodplains, rapidly grows tall and slender. But on gravelly upland soil, it grows more slowly and branches out, becoming a round-topped tree. Kentucky Coffeetree is rarely seen in the forest. It has a wide range but is common nowhere.

Kentucky Coffeetree's remarkably open frame is visible long after other trees have leafed out in the spring. The contorted twigs appear to be lifeless and without buds; on closer inspection, two or three dark spots can be seen in a row just above the large leaf scar. These are the tips of tiny buds tucked away in crater-like depressions. By the end of June, when the leaves are fully grown, it seems impossible that such size could have been concealed in so tiny a bud. The leaves of Kentucky Coffeetree are the largest of the woodland trees, often growing 2 to 3 feet long and half as wide. They are twice compound leaves, giving the tree's foliage a tropical look.

The flowers of Kentucky Coffeetree are small, held in greenish-white clusters. Male and female flowers are usually found on separate trees. The seeds are in large, thick pods that remain on the tree throughout winter, looking like blackbirds among the branches. The seeds are round, smooth, and hard to crack. The tree's name comes from its kernel, which was roasted and used by early Kentucky settlers as a coffee substitute.

The bark of Kentucky Coffeetree has an unusual gray color, is deeply furrowed, and has long wave-like ridges up and down the trunk.

Gymnocladus dioicus · Kentucky Coffeetree

habitat . . .
FOREST. lowland wet-mesic, floodplain and farmstead planting
- zone – 5a

form . . .
IRREGULAR. ovoid, open large canopy tree (75–100′)
- branching – coarse to sparse LATERAL limbs
- twig – stout, brown, blunt-ended with thick brown pith having MINUTE black buds on large leaf scar
- bark – gray-brown, shallow-furrowed with narrow, VERTICAL ridges; CURLING toward one side

foliage . . .
ALTERNATE. bipinnately COMPOUND, leaflets (30–76) oblong-OVATE, dark blue-green (1–3″) with entire margin; pale beneath, pink unfolding
- color (fall) lemon-YELLOW
- season – deciduous

flower . . .
SPIKE. large, yellow-GREEN, hairy, (3/8–5/8″) open, UPRIGHT clusters; terminal
- sex – monoecious

fruit . . .
LEGUME. hard-shelled, RED-brown, BEAN-like pod (4–7″) with many rounded, shiny brown seeds
- season – maturing late autumn

MIXED FLOODPLAIN COMMUNITY

Butternut is a medium-sized tree that usually grows in moist, rich soils in floodplains. It is also found scattered throughout upland woods. It is so common in the New England states that its nuts serve as a substitute for black walnuts.

The name Butternut comes from the mellow flavor of the nutmeat, which is considered desirable by many. Butternut's light gray bark with flat ridges makes it easy to identify when the tree is mixed with the other hardwoods.

Butternut is much smaller in height than Black Walnut. Its stems are much more branched, and it has a more scraggly appearance. The leaves have longer stems with fewer leaflets, so the tree looks sparser. Butternut wood is lightweight and easily worked. It is used in fine carvings and church altars where it seems to mellow with age.

Butternut is a fast-growing tree, but only a few trees survive beyond seventy-five years of age.

habitat . . .
FOREST. upland mesic, lowland wet-mesic, stream terraces and rock slopes
- zone–3a

form . . .
IRREGULAR. globular, open small canopy tree (50–75′)
- branching–large, ASCENDING limbs on straight trunk
- twig–stout, VELVETY, tan-brown with chambered pith having WOOLLY, conical bud on large leaf scar; tan-brown
- bark–smooth, ash gray, SILVERY, shallow-furrowed with FLAT-wide ridges

foliage . . .
ALTERNATE. pinnately COMPOUND, leaflets (11–17) oblong-LANCEOLATE, yellow-green (2–4½″) with fine-toothed margin; pointed tip, smooth above, tomentose beneath
- color (fall) YELLOW-gold
- season–deciduous

flower . . .
CATKIN. cylindrical, yellow-GREEN, slender (3–4″) drooping clusters; (6–8) at tip of twig; terminal
- sex–monoecious

fruit . . .
NUT. elliptic, tan-brown, FOOTBALL-shape (2″) with FUZZY husk in clusters of 3–5; drooping, having EDIBLE seed
- season–maturing late autumn

MIXED FLOODPLAIN COMMUNITY

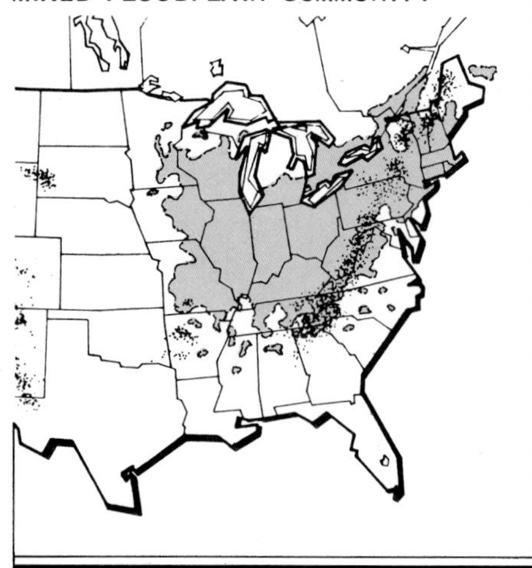

Juglans cinerea · Butternut

Eastern Black Walnut is a rugged tree that grows on floodplain terraces. It was thoughtlessly destroyed in land-clearing operations by early settlers, only later to be in great demand for its beautiful wood for cabinet and furniture making.

Black Walnut can be a tall, upright giant in the woods or a rather symmetrical, round-topped tree when planted as a specimen. At one time, there were trees 150 feet in height with trunks 8 feet in diameter, but it is hard to find one that is even 100 feet tall today.

The leaves are compound with large leaflets giving the whole tree a fern-like appearance by midsummer. The fruit of Eastern Black Walnut is perhaps the best known of the native nut trees. It has become famous for the flavor it imparts to cookies, cakes, and breads.

If Black Walnut trees are planted too close to a house, the nuts may become a nuisance because they litter, stain, and make noise as they tumble down the roof.

habitat . . .
FOREST, SAVANNA. upland mesic-dry, lowland wet mesic, floodplain, stream edge and farmstead planting
- zone – 4b

form . . .
IRREGULAR. ovoid, open large canopy tree (75–100′)
- branching – stout, ASCENDING limbs
- twig – gray, tan-brown, VELVETY, stout with chambered pith having blunt, WOOLLY buds on large leaf scar; tan-brown, gray
- bark – thick, brown-black, deeply FURROWED with rectilinear ridges, straight trunk

foliage . . .
ALTERNATE. pinnately COMPOUND, leaflets (15–32) oblong LANCEOLATE, yellow-green (2½–5″) with fine-toothed margin; glabrous above, hairy beneath
- color (fall) YELLOW-gold
- season – deciduous

flower . . .
CATKIN. cylindrical, yellow-GREEN, pencil-like (4″) drooping clusters
- sex – monoecious

fruit . . .
NUT. yellow-green, ball-shaped, (1½–2½″) dia. in clusters of 2 or 3; HUSK smooth with EDIBLE seeds
- season – maturing late summer

MIXED FLOODPLAIN COMMUNITY

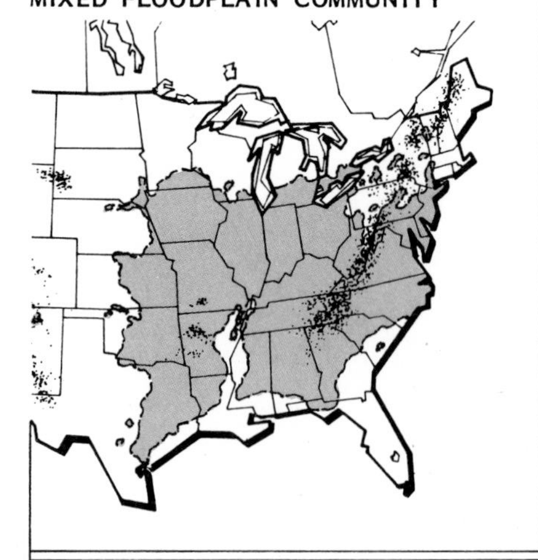

Juglans nigra · Eastern Black Walnut

American Elm has for generations been the best-loved tree in a region stretching from New England west to the Rocky Mountains and south to Texas. Its graceful, vase-shaped form lined the streets of almost every town in the Midwest. Street widening projects that would cause the removal of Elms were often fought and defeated with ease. Then Dutch Elm disease gained a foothold in the middle Atlantic states and gradually moved west, taking with it almost every giant Elm in its path.

Today, cities plant a mixture of tree varieties, though none has really replaced the grand stature of the Elm. However, when a new tree disease sweeps the nation, it won't leave whole towns without a shade tree remaining.

Not all American Elms died from Dutch Elm disease. Some were treated in time; others somehow escaped on their own. But even though most mature Elms were killed, we will always have Elms. This tree is too prolific in seed production to ever die out completely.

habitat . . .
FOREST, SAVANNA. upland mesic, upland mesic-dry, lowland wet-mesic, floodplain, ravines, slopes and open fields
- zone – 2

form . . .
GLOBULAR. vase-shaped, large canopy tree (75–100′)
- branching – open, DROOPING limbs
- twig – slender, red-brown, ZIG-ZAG with conspicuous lenticels having lateral, long-pointed red buds; brown margin
- bark – thick, gray-brown, deep-furrowed with narrow ridges of BLEACHED appearance

foliage . . .
ALTERNATE. simple, oblong-OBOVATE bright green leaves (1¼–2½″) with DOUBLE-toothed margin; uneven base, thick, rough above
- color (fall) YELLOW
- season – deciduous

flower . . .
CLUSTER. small, red-brown, BELL-shaped (⅛″) with slender STRING filaments
- sex – monoecious

fruit . . .
SAMARA. flat, tan-brown, elliptic, (⅜″) dia., WAFER-like, one-seeded key; hairy, 2-pointed, NOTCHED apex
- season – maturing early spring

MIXED FLOODPLAIN COMMUNITY

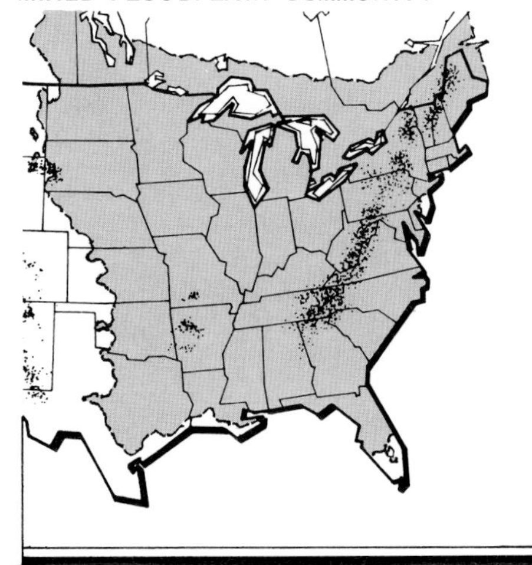

Ulmus americana · American Elm

Slippery Elm is a small-sized canopy tree that reaches 50 to 70 feet in height. It prefers moist soils, especially the lower slopes of floodplains, but it is also found in dry uplands mixed with other hardwood trees.

Slippery Elm is not affected by Dutch Elm disease, but it has many characteristics like those of American Elm. Its leaves are similar except much larger and more roughly textured on the upper surface. The flowers are small clusters like American Elm's, but they are red rather than greenish. The small, wafer-like seed clusters that follow are also similar, but they are not as deeply notched at the tip nor do they have the hairy wing edge.

Slippery Elm has large, prominent winter buds that are covered with a rust-red fuzz. The bark is much lighter gray and has an inner bark that is slippery and gummy. This inner bark was once chewed by small children as a popular pastime; now an extract from it is available through drugstores in the form of cough medicine and throat lozenges.

Even though in many ways Slippery Elm is similar to its once-popular counterpart, it should not be considered a true replacement for American Elm, because it lacks that tree's height and vase-like form.

habitat . . .
FOREST, SAVANNA. upland mesic-dry, lowland wet-mesic, floodplain, ravine slopes and abandoned fields
- zone – 4

form . . .
GLOBULAR. irregular, open, large canopy tree (75–100′)
- branching – large, spreading, ASCENDING limbs
- twig – stout, ash gray, hairy with blunt, hairy buds; dark brown-black
- bark – light brown, deep-furrowed with inner bark SLIPPERY (mucilaginous)

foliage . . .
ALTERNATE. simple, elliptic-OBOVATE dark green leaves (2–3″) with DOUBLE-toothed margin; uneven base, long-pointed, thick with parallel side veins, very ROUGH above, hairy beneath
- color (fall) dull YELLOW
- season – deciduous

flower . . .
CLUSTER. yellow-green, small, (1/8–1/4″) dia. in dense clusters
- sex – monoecious

fruit . . .
SAMARA. nearly flat, tan-brown WAFER-like, (1/2–3/4″) dia., one-seeded key; hairless, slightly notched
- season – maturing early spring

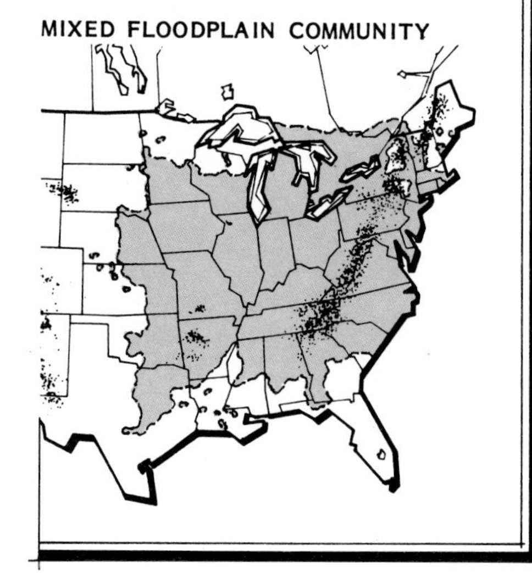

Ulmus fulva · Slippery Elm

Eastern Redbud matures to a 20- to 40-foot, round-topped tree, often resembling an apple tree. Only a few trees can provide a more useful and efficient form for defining a room-like space outdoors. With continual removal of the dead interior branches, both apple and redbud provide a perfect umbrella of leaves. These trees give people the ever-elusive but desirable feeling of being inside when they are actually outside.

My own family's choice of homes twenty-five years ago was greatly influenced by the presence of one Redbud tree near the back of the lot. This is where I chose to develop our "outdoor living room." It is a successful space largely because of the Redbud, which not only provides the ceiling but gives us a wealth of extras.

Each spring this Redbud blooms in gay profusion, covering its branches with pea-like, rose-colored to magenta blossoms. In summer the heart-shaped leaves fill in the space to shade and cool the terrace below. These same leaves turn a lovely yellow-green in autumn, providing a wall of color when viewed from the house. And all winter, I enjoy the angular, sculptural branching that is so common to Eastern Redbud.

Unfortunately, I have had to string numerous cables to prevent the tree from spreading so wide as to damage itself. Also, because of the average life span of Redbud, I will be lucky to have this tree another twenty-five years.

habitat . . .
FOREST. upland dry, lowland wet-mesic, stream edge, slopes and rocky bluffs
- zone—5a

form . . .
GLOBULAR. broad, small understory tree (20–35′)
- branching—spreading, PICTURESQUE limbs
- twig—glossy, red-brown, ZIG-ZAG having buds on leaf scar with 3-BUNDLE scars; glossy black
- bark—thin, red-brown, MAROON with scaly plates, easily EXFOLIATING

foliage . . .
ALTERNATE. simple, CORDATE (broad to HEART-shaped), dull green leaves (2½–4½″) with entire margin; long-stalked, glabrous (young)
- color (fall) golden YELLOW
- season—deciduous

flower . . .
CLUSTER. small, pink, (½″), PEA-like in dense clusters along stem
- sex—dioecious

fruit . . .
LEGUME. flat, purple-brown, bean-like (2″), PAPERY pod; clusters along stem
- season—maturing late summer, PERSISTING through winter

MIXED FLOODPLAIN COMMUNITY

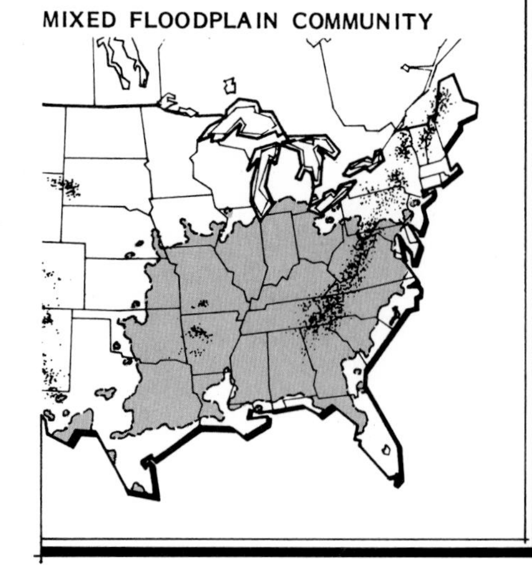

Cercis canadensis · Eastern Redbud

Eastern Wahoo, or Burning Bush, is a large shrub. If not crowded by other plants, it can develop into a small tree 20 to 25 feet tall. It is common in low, wet areas at the base of mesic slopes, in alluvial floodplains, and bordering small streams.

Wahoo is an Indian name. The plant's bitter leaves and bark were known as good medicine, but not much use is made of them by modern science. The English call this plant Spindle-tree, because its wood has long been used to make spindles and knitting needles. Toothpicks and skewers are also made from its wood (Werthner, 1935).

The bark on Eastern Wahoo's twigs is greenish-gray and is bordered by corky lines that make the stems appear four-sided. The leaves are large and resemble those of American Plum, except that Wahoo leaves are opposite along the stem while American Plum's are alternate.

The flowers of Eastern Wahoo are small with four purple petals, resembling a Maltese cross. They are succeeded by small, spectacular fruits that are the most prominent attribute of this delicate, little tree. As Thoreau said, "its fruit is its flower" (Rogers, 1931). The brilliant red fruit is first displayed in autumn. The crimson pods split to expose their scarlet-coated seeds, which hang on the plant far into winter, brightening an otherwise colorless landscape.

habitat . . .
FOREST. lowland wet-mesic, floodplain, stream and ravine edge
- zone—4b

form . . .
GLOBULAR. irregular, open, small understory tree (20–35′)
- branching—stiff, UPRIGHT, spreading limbs
- twig—lime-green, brown, slender, 4-LINED (winged) with small purple-green buds
- bark—smooth, ash gray-green with thin, flaky scales

foliage . . .
OPPOSITE. simple, oblong-OVATE green leaves (2–4″) with fine-toothed margin; base wedge-shaped having prominent veins
- color (fall) PINK-bright red
- season—deciduous

flower . . .
CLUSTER. small, red-purple, (3/8″) wide with 4 petioles on underside of twig; wide forking stalks
- sex—monoecious

fruit . . .
CAPSULE. small, flesh pink, CHAMBERED, (5/8″) wide, 3–4 opening in drooping clusters; seeds GLOSSY RED
- season—maturing in autumn

MIXED FLOODPLAIN COMMUNITY

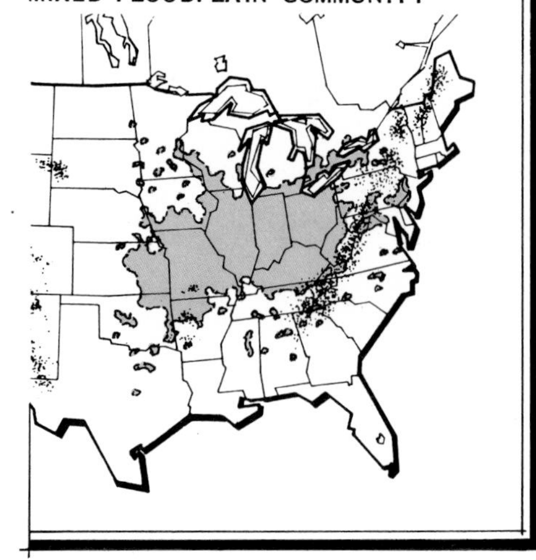

Euonymus atropurpureus · Eastern Wahoo

When the Red Mulberry was discovered in great abundance in Virginia, it was thought to be the answer to a new industry in the colonies, a replacement for the **White Mulberry** of China. Most trees are valued for their wood or fruit, but White Mulberry is valued for its leaves, which are fed to silkworms. However, the native Red Mulberry did not succeed as a replacement for White Mulberry, so the Chinese version was introduced (Rogers, 1931).

Mulberries belong to a family of trees that exude a milky sap from broken leaf stalks. This family also includes Osageorange. Red Mulberry has a short, leaning trunk and a low, wide-spreading crown that casts dense shade. Its leaves are thick, with no two being alike. Some have a lobe or two; others develop so many that they turn into fancy designs. Red Mulberry and its cousin White Mulberry prefer the deep, rich, moist soils in floodplains, but they are also tolerant of thin soils and rocky slopes.

The fruit of the Red Mulberry normally matures when it turns from red to black. White Mulberry fruit normally matures when it turns from green to white, but in the Midwest they have hybridized so completely that it may bear red and black fruit like its cousin.

Morus rubra · Red Mulberry

Morus alba · White Mulberry

habitat . . .
FORESTS, SAVANNA. upland dry, lowland wet-mesic, floodplain, creek edge, open hills and pastures
- zone—5b

form . . .
GLOBULAR. broad, large understory tree (35–50′)
- branching—stout, dense, ASCENDING limbs
- twig—smooth, ORANGE-brown, slender with glossy buds; light brown
- bark—orange-brown, scaly with raised LENTICELS on furrowed, FIBROUS ridges

foliage . . .
ALTERNATE. simple, CORDATE dark green leaves (4–7″) with COARSE-toothed margin; tomentose, occasionally LOBED (mitten-shape)
- color (fall) golden YELLOW
- season—deciduous

flower . . .
CATKIN. yellow-green, elliptic (1″), dense clusters
- sex—dioecious

fruit . . .
BERRY. dark red-PURPLE, cylindric (1″) with many one-seeded fruits; EDIBLE
- season—maturing early summer

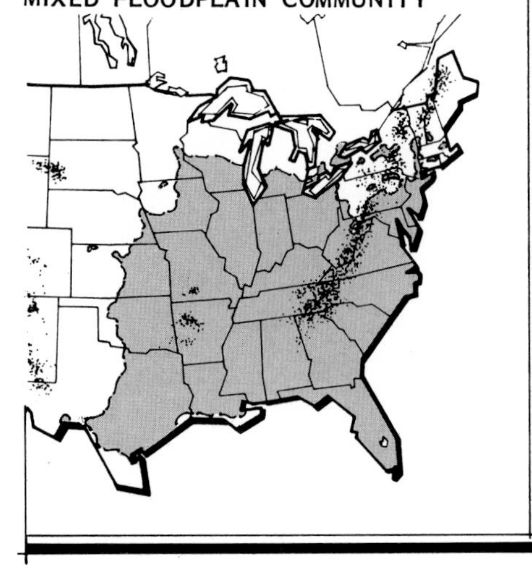

American Bladdernut is a small tree or large shrub, 8 to 15 feet high, that is found along the borders of damp woods, floodplains, and bottomlands. The plant grows in any good garden soil and improves greatly under cultivation; it flowers more abundantly, grows more luxuriantly, and becomes more symmetrical than those growing in the wild.

American Bladdernut has small, bell-shaped flowers that are not very attractive individually, but when the full-flowered, drooping racemes cover the bush, the effect is very appealing. The plant also has unusual trifoliate leaves. Its greenish bark is textured with linear white cracks, but the most interesting feature is its fruit.

The fruit appears late in summer. It is first green; then, as the leaves yellow, the fruit changes to brown. As the name implies, the fruit is a small bladder with a nut inside. Each bladder has three inflated chambers. In each chamber or cell, there are usually one or two seeds that mature, break loose, and rattle about when the pod is shaken. Those of a European species are often strung together as rosary beads.

American Bladdernut is a fast-growing shrub that should be used as an ornamental. Unfortunately it is not often available from nurseries.

habitat . . .
FOREST, SAVANNA. upland mesic-dry, lowland wet-mesic, stream and ravine edge
- zone—3

form . . .
OVOID. upright, small understory tree (12–30′)
- branching—erect, often vase-like appearance
- twig—slender, gray-brown with white STRIPES or streaks having small, 2-scaled buds; brown
- bark—GRAY-GREEN with wart-like LENTICELS; narrow, white pith

foliage . . .
OPPOSITE. pinnately COMPOUND, leaflets (3–5) elliptic-OBOVATE, yellow-green leaves (1½–3½″) with margin unevenly toothed; tomentose beneath
- color (fall) YELLOW-green
- season—deciduous

flower . . .
CLUSTER. lemon-YELLOW, small (½″), BELL-SHAPED in drooping (4″) clusters; terminal
- sex—monoecious

fruit . . .
CAPSULE. tan-brown, PAPERY, 3-lobed, (1–3″) inflated pod; shiny brown seeds
- season—maturing late summer

MIXED FLOODPLAIN COMMUNITY

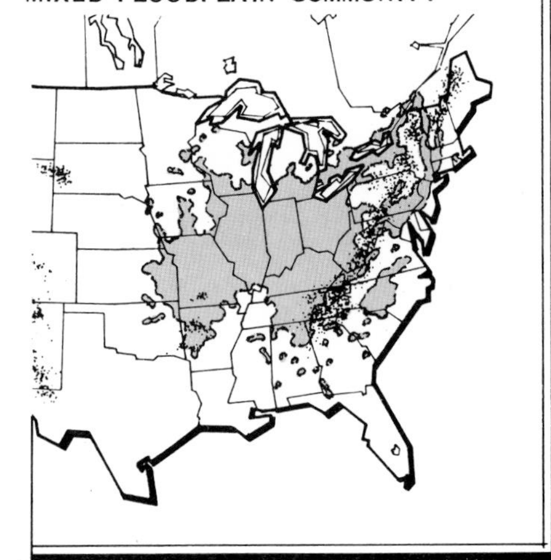

Staphylea trifolia · American Bladdernut

American Elder, or Elderberry, is a tall shrub that never quite made it as an ornamental, probably because some people consider it too common. It grows in floodplains, along streams, and in fencerows.

Birds delight in this plant, for in September the branches become so heavily laden with small, black berries that the stems often break. The berries are full of juice, "too bitter to be eaten raw (except by birds) and too sweet to jell without adding pectin" (Borland, 1983). However, these berries are just right for making wine.

American Elder flowers bloom in late spring and have large, attractive, round-headed clusters of small, creamy white blossoms. The leaves, which are compound, have no fall color. When they fall, they leave the shrub quite bare. "The branches are pithy and for many years they were used to make taps for collecting sap for maple sugar production" (Borland, 1983).

Even though American Elder is not as popular as it could be, it should be considered for hedges, thickets, and even specimen plantings, for it continually rejuvenates itself with new growth emerging at the base of dead branches.

Sambucus canadensis · American Elder

habitat . . .
FOREST, SAVANNA. upland mesic-dry, lowland wet-mesic, stream and water edge, open fields and roadsides
- zone—4

form . . .
OBOVOID. upright, large shrub (6–15′)
- branching—stiff, ERECT to arching limbs
- twig—WOODY, tan-brown, stout; SWOLLEN at joints with green-brown buds
- bark—yellow-tan, brown with WART-like lenticels, having thick, WHITE pith

foliage . . .
OPPOSITE. pinnately COMPOUND, leaflets (5–11) ELLIPTIC-ovate, yellow-green leaves (1½–3½″) with coarse-toothed margin; veins tomentose beneath
- color (fall) YELLOW-green
- season—deciduous

flower . . .
CLUSTER. showy, creamy WHITE, small (¼″), FLAT-topped (4″) across, upright clusters
- sex—monoecious

fruit . . .
BERRY. small, PURPLE-black, (¼″) dia. in flat-topped (4″) across clusters
- season—maturing in summer, early autumn

MIXED FLOODPLAIN COMMUNITY

Arrowwood Viburnum, or Southern Arrowwood, is an erect shrub, 3 to 6 feet tall, with slender, smooth, gray-barked stems. It closely resembles the Rafinesque Viburnum of the north but may be distinguished by its hairy branchlets and lower leaf surfaces, and longer leaf stems.

Typical of the Viburnum Family, Arrowwood is a colorful plant throughout the seasons. In June, it is covered with flat clusters of snowy white flowers. These are followed by dark, shiny, blue berries.

When the berries mature, the plant begins its most attractive period. The combination of dark berry clusters and shiny, coarse-toothed leaves with their prominent veins is quite striking. The plant's appearance becomes even more striking when the leaf color deepens in autumn to a wine-red, producing an elegant combination of red leaf and blue berry. The berries are then eaten by birds, although it is difficult to understand why, because the berries are dry, tasteless, and rather seedy.

Arrowwood Viburnum is hardy in the northern states as well as the southern states, and it is now available from many nurseries. It is highly recommended for home landscaping and for park plantings because of its attractiveness throughout the seasons.

habitat . . .
FOREST, SAVANNA. upland mesic-dry, lowland wet-mesic, alluvial flat, stream edge and open slopes
- zone—2

form . . .
GLOBULAR. erect, large shrub (6–15′)
- branching—straight, ASCENDING limbs
- twig—ridged, gray with bright yellow stems and scaly buds; red-brown
- bark—SMOOTH, brown

foliage . . .
OPPOSITE. simple, ovate, green leaves (1½–3½″) with COARSE-toothed margin; hairy beneath, straight, PROMINENT veins
- color (fall) red-MAROON
- season—deciduous

flower . . .
CLUSTER. dense, creamy white (2″) across FLAT-TOPPED, upright clusters; profuse
- sex—monoecious

fruit . . .
BERRY. elliptic, BLUE-black (¼–⅜″), flat-topped clusters
- season—maturing summer, early autumn

MIXED FLOODPLAIN COMMUNITY

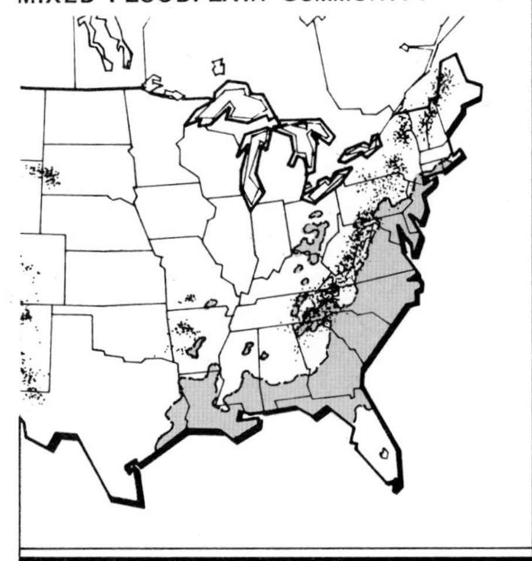

Viburnum dentatum · *Arrowwood Viburnum*

Virginia Creeper is one of several vines in the floodplain, and it is by far the most versatile. It trails across the ground as a ground cover and scrambles over shrubs and thickets. In wooded areas, it climbs to the top of the tallest tree. Its stems cling steadfastly by means of strong tendrils that are tipped with adhesive disks. These disks make Virginia Creeper adaptable in cities as well, where the plant clings to wood siding as well as stone and brick masonry. Once established, it is very vigorous and tolerates sun or shade.

In the shade, the five large leaflets develop only a pale yellow fall color. But in the tops of trees or on the sunny sides of buildings, the vine becomes an attractive deep red in the fall.

The flowers are quite inconspicuous, appearing in June or July as small, greenish blossoms. The blue-black berries that follow and ripen in September are showy after the leaves have fallen and are eaten with great relish by many species of wild birds. By human standards, however, the berries are far from edible.

Virginia Creeper is a useful vine for covering nearly anything in the landscape from bare ground to fences and entire buildings. Its only drawback is its own vigor: it can become a nuisance because it shows no preference for walls over window openings.

habitat . . .
FORESTT, SAVANNA. upland mesic, lowland wet-mesic, mesic-dry, woods edge, ravines and bluffs
- zone – 2

form . . .
HIGH. climbing vine with dense branching form (3–40′)
- twig – REDDISH stems with tendrils branched by SUCTION discs and small buds; red-brown
- bark – brown, SCALY to shallow furrowed

foliage . . .
ALTERNATE. palmately COMPOUND, leaflets (5) oblong OVATE-lanceolate, dark green leaves (2½–5″) with COARSE-toothed margin; stems to 8″ long
- color (fall) SCARLET-red
- season – deciduous

flower . . .
CLUSTER. small, YELLOW-green (⅛–¼″), open (inconspicuous) branched clusters
- sex – monoecious, appearing after leaves

fruit . . .
BERRY. small, blue-BLACK (¼″), globular, open (4″) dia. clusters
- season – maturing late autumn, PERSISTING through winter

MIXED FLOODPLAIN COMMUNITY

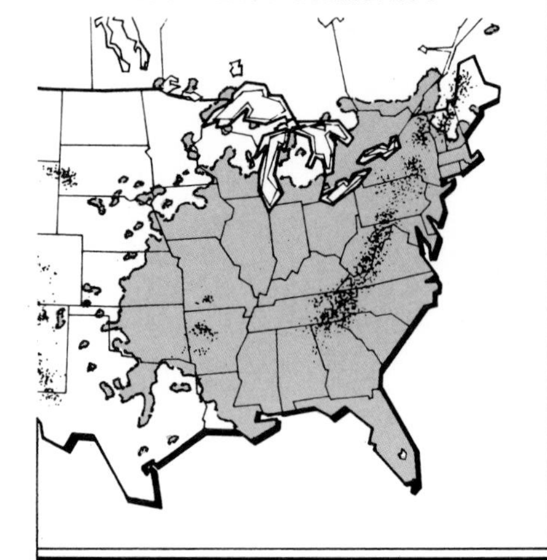

Parthenocissus quinquefolia · Virginia Creeper

Common Moonseed is a vine that grows vigorously but does not climb high. It is common in floodplains and climbs on trees and shrubs by twining around small branches. It usually dies back to the ground each winter.

Common Moonseed blooms in May and June with tiny, greenish-white blossoms that are rather inconspicuous. The fruit that follows, maturing in September, is poisonous. Common Moonseed can be mistaken for Wild Grape, which may account for the many cases of poisoning that are attributed to eating wild grapes in the woods.

To distinguish Moonseed vine from Wild Grape, look for leaves that are not toothed or smash a berry and look for a flattened, crescent-shaped seed.

habitat . . .
FOREST. lowland wet-mesic, mesic-dry, ravines, woods edge and steep rocky bluffs
- zone – 2

form . . .
LOW. climbing vine often TWINING with stems, hardly woody (3–15′)
- twig – GLOSSY green-tan brown with bud HIDDEN by large leaf scar
- bark – smooth, brown; often becoming WARTY (old)

foliage . . .
ALTERNATE. simple, ovate-CORDATE, (3–7) lobed, dark green leaves (3–5″) with entire margin; 3–7 main venation, tomentose beneath
- color (fall) YELLOW
- season – deciduous

flower . . .
CLUSTERS. small, yellowish WHITE (¼–⅜″), slender, open clusters; drooping
- sex – dioecious, appearing after leaves

fruit . . .
BERRY. small, blue-BLACK (⅜″), grape-like clusters with flat to concave, CRESCENT, one seed; MOON-shaped
- season – maturing late summer

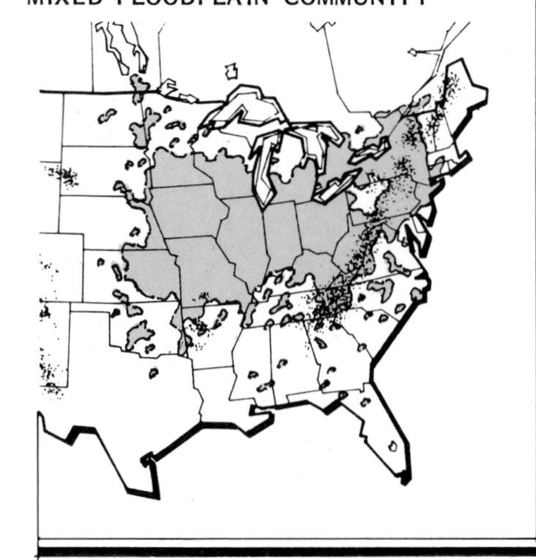

habitat . . .
FOREST. lowland, wet-mesic, mesic, mesic-dry, woodland stream banks, ridges and thicket clearings
•zone—5a

form . . .
LOW. climbing vine often forming (impenetrable) thickets (3–12′)
•twig—ZIG-ZAG, green-brown stem, often 4-sided with prickles; small buds HIDDEN by leaf base; smooth
•bark—smooth, dark green, ARMED with many spur-like SPINES; black to purple

foliage . . .
ALTERNATE. simple, ovate-CORDATE, dark green leaves (2–6″) with entire, wavy margin; apex sharp-pointed, palmately VEINED (3–5) with tendrils at base of petiole
•color (fall) YELLOW-brown
•sex—deciduous

flower . . .
CLUSTER. small, yellow-green (1/8–1/4″), open clusters; PROFUSE
•sex—dioecious, appearing after leaves

fruit . . .
BERRY. small, black with BLUISH bloom (1/8–1/4″), globular dense cluster
•season—maturing late summer, PERSISTING

MIXED FLOODPLAIN COMMUNITY

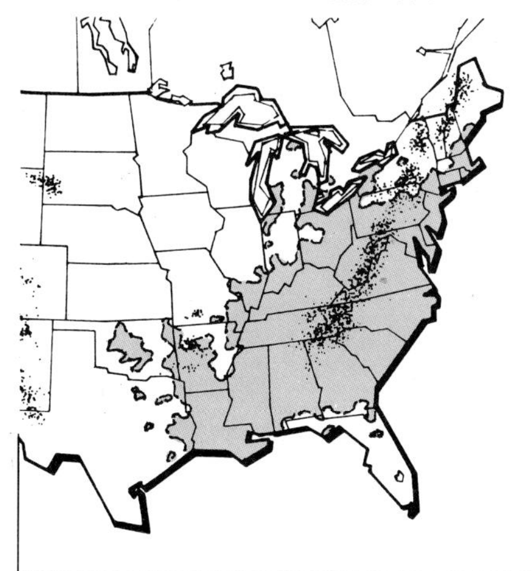

Common Greenbriar is a vigorous, thorny vine that grows with long, thin tendrils. It frequently forms impenetrable and formidable tangles and often climbs to 30 feet in trees and shrubs.

The leaves of Common Greenbriar are spade-shaped, having parallel veins that start at the base of the leaf and end at the tip. The flowers are small and greenish-white. They bloom in June and mature in September to a black cluster of grape-sized fruit. The fruit is eaten by many species of wild birds, including Ruffed Grouse, Wild Turkey, and Ring-necked Pheasant.

Common Greenbriar is a common vine in floodplains. It prefers low areas but also grows on drier sites and upland slopes.

Smilax rotundifolia · Common Greenbriar

Poison Ivy can be a groundcover, a shrub-like plant, or a vine climbing by aerial rootlets in the tops of tall trees. All parts of Poison Ivy contain a poisonous, oily substance that blisters the skin, causing extreme itching and burning.

The leaves are grouped in threes, varying in shape but usually notched near the tip. They turn beautiful shades of orange, yellow, and red in the fall, providing some of the first color in the floodplain. The support vines are densely covered with aerial rootlets, and the fruit is a small, waxy, white berry. Birds are quite fond of the loose clusters of berries.

habitat . . .
FOREST, SAVANNA. upland mesic, lowland wet-mesic, mesic, open woodlands, floodplain and stream edge
- zone – 3a

form . . .
LOW. groundcover or climbing vine, often ERECT branching form (3–40′)
- twig – VELVETY gray-grown, flexible to smooth with WOOLLY, yellowish red buds; large leaf SCAR
- bark – smooth, REDDISH brown covered with FIBROUS aerial rootless; (old) slightly fissured

foliage . . .
ALTERNATE. trifoliate COMPOUND, leaflets (3) asymmetric ovate-lanceolate, glossy green leaves (5–12′) with entire coarse-toothed margin; smooth and pale beneath
- color (fall) reddish ORANGE-scarlet
- season – deciduous

flower . . .
SPIKE. small, GREENISH white (1/8–1/4″), (inconspicuous) axillary (1/4″) cluster spikes; dense
- sex – dioecious

fruit . . .
BERRY. small, WHITE-ivory (1/8–1/4″), loose clusters up to 8″; drooping
- season – maturing late summer, PERSISTING

MIXED FLOODPLAIN COMMUNITY

Toxicodendron radicans · Poison Ivy

habitat . . .
FOREST, SAVANNA. lowland wet, lowland wet-mesic, mesic-dry, floodplains, woods edge, ravines and stream banks
•zone – 2

form . . .
HIGH. dense, climbing vine, open at base (3–45′)
•twig – flexible, REDDISH green becoming tan-brown; slightly ridged with 2-scale, conical bud
•bark – EXFOLIATING, orange-brown in thin, linear, stringy plates

foliage . . .
ALTERNATE. simple, palmately 3-LOBED, CORDATE glossy green leaves (3–7″) with coarse-toothed margin; deep MAPLE-like sinuses; thin, shiny beneath
•color (fall) YELLOW-green
•season – deciduous

flower . . .
SPIKE. small, YELLOWISH green (3/8–1/2″), open (inconspicuous) flower cluster
•sex – dioecious, appearing after leaves

fruit . . .
BERRY. small, PURPLE to black (1/8–1/4″), bluish bloom in dense PYRAMIDAL clusters
•season – maturing early summer

MIXED FLOODPLAIN COMMUNITY

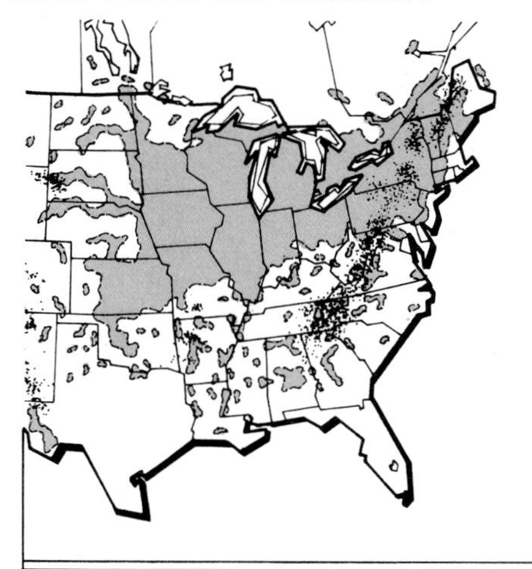

Wild Grape is found on rich, alluvial bottomlands where it climbs by tendrils, sometimes overlapping small trees and shrubs.

This vine creates nesting sites for cardinals and catbirds (Degraaf and Whitmann, 1979). Wild Grape is among the most valuable foods for wildlife. Its fruit is juicy and tart and entirely too sour to be eaten raw, although it makes good jelly, preserves, and wine.

The grape leaf resembles a sycamore leaf, because it is large and coarse-toothed. Old vines of Wild Grape can be quite large. They have shredded bark and often hang from trees like ropes that Tarzan might use, giving the floodplain a jungle-like appearance.

Vitis species · Wild Grape

Willow, maple, and poplar are common to the River–Lake Margin community. This community follows water courses and is characteristically linear, like a green ribbon weaving itself through the rural fabric of patterned fields. The most visual feature, typical of this river (riparian) community is the distinct bands of vegetation of different heights; often there are willows next to the water, taller soft maples behind them, and giant poplars taking up the background.

Spring flooding is common to this community, delaying the growth of ground cover plants by prolonged submersion. Plants in this community are, by necessity, extremely tolerant of flooding and capable of withstanding several days of inundation.

The Lake Margin community is not as dramatic in its plant expression. This community lacks the dynamic changing qualities associated with the river. In contrast to the river, the lake margin appears to remain, for the most part, placid and austere. But lake levels do fluctuate, thus making demands on the plants at the water's edge.

The River and Lake Margin communities have in common plants that can tolerate change. These plants even seem to prefer wet soils where others cannot grow because of the lack of oxygen in the soil.

River-Lake Margin Community

Trees: Canopy (over 48′)

Shrubs: (6-12′)

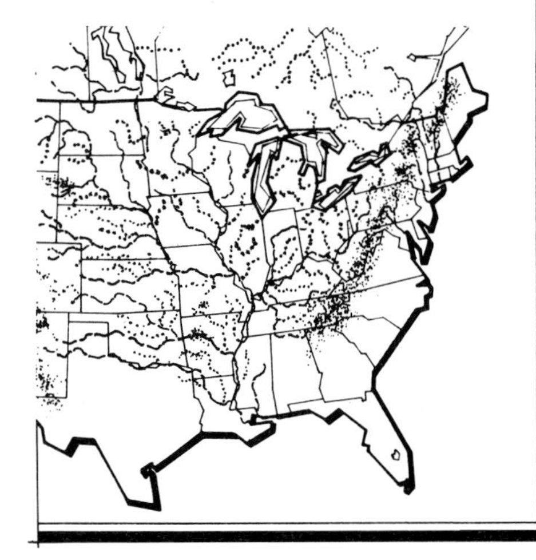

Boxelder is in the Maple Family, but it doesn't look like other maples because of its irregular growth and compound leaves. It is a rapidly growing, moderately tall tree that is found along streams and rivers and is tolerant enough to withstand unfavorable city conditions. In winter, young Boxelder saplings and mature specimens alike are easy to recognize by their twigs. The new twig growth is maroon and partially covered with a whitish film. The winged samaras, each containing a fat nut-like seed, are borne in bunches that hang like tan tassels. Once the leaves have fallen, the samaras provide a nutritious handout to hungry birds and squirrels. Boxelder is considered the least attractive of the large maples because of its brushy crown and sprouting base and because its soft wood is subject to snow and wind damage. In addition, Boxelder lives only a relatively short time.

habitat . . .
FOREST, SAVANNA. lowland wet, lowland wet-mesic, stream-lake margin and ravine edge
- zone – 2

form . . .
OBOVOID. irregular, large understory tree (35–50′)
- branching – stout, ascending limbs on short, DIVIDING trunk
- twig – slender, greenish PURPLE, glabrous with woolly, reddish blunt buds; silvery leaf scar ENCIRCLING twig
- bark – thin, gray-brown, blocky to DEEP-furrowed with broad ridges

foliage . . .
OPPOSITE. simple, pinnately COMPOUND, leaflets (3–7) ovate-lanceolate, yellow-green leaves (3–4″) with coarse-toothed margin; long-pointed tip, tomentose beneath
- color (fall) YELLOW-green
- season – deciduous

flower . . .
CLUSTER. (male) RED (1/8″); (female) yellowish green, dense clusters on long, WIRE-like filaments; pendulous
- sex – dioecious

fruit . . .
SAMARA. tan-brown, paired WING (2″) keys in dense clusters; terminal
- season – maturing in summer, PERSISTING through winter

RIVER-LAKE MARGIN COMMUNITY

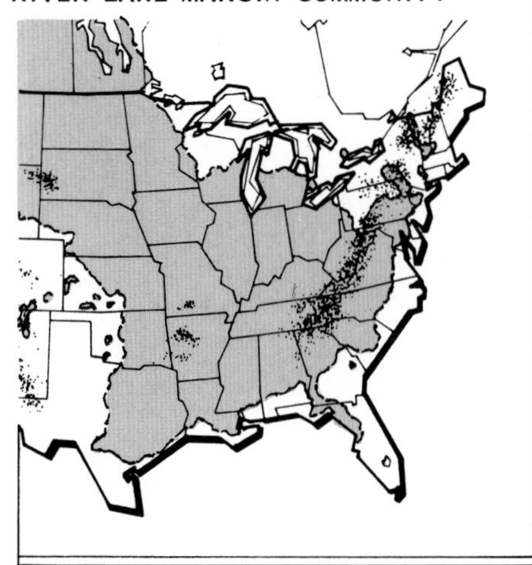

Acer negundo · Boxelder

Red Maple is perhaps more deserving of its name than any other tree. In earliest spring, the buds of Red Maple begin to turn from their dull maroon winter color to a brighter, lighter red hue. When the blossoms open, they are also red. The same red is found in the young leaves and the winged seeds, and it tinges the twigs and leaf stems all summer. In fall, Red Maple is in perhaps its most brilliant form, with its startlingly red leaves. Even when winter comes, the red tone is not completely lost. Instead it lingers in the younger bark and buds.

Red Maple is a fast-growing soft maple that drops its leaves early in the season. It is a handsome tree but less symmetrical than the hard maples.

Red Maple is a good shade tree. As with other maples, however, it is difficult to keep a lawn growing within its shade after the tree has reached maturity.

Red Maple grows along streams, rivers, and wet meadows. It can live 150 years.

habitat . . .
FOREST. upland dry, lowland wet, lowland wet-mesic, steep rocky slopes, and low wet areas
- zone – 3a

form . . .
OVOID. globular, large canopy tree (75–100′)
- branching – stout, ASCENDING limbs
- twigs – slender, bright RED, lustrous with clustered (1/4″) bright RED buds; blunt
- bark – smooth, SILVERY-gray (young) with long, narrow, SCALY plates; exfoliating (old)

foliage . . .
OPPOSITE. simple, palmately 3-LOBED, light green leaves (2½–4″), toothed margin with sharp, ANGLED sinuses; long petiole with silvery-gray or whitened, tomentose beneath
- color (fall) scarlet, CRIMSON-red
- season – deciduous

flower . . .
CLUSTER. small, RED (3/8″), dense clusters with short, SPIDER-like filaments
- sex – dioecious/monoecious

fruit . . .
SAMARA. red-reddish brown (3/4–1¼″), paired KEYS with wide divergent WINGS; dense terminal clusters
- season – maturing in spring

RIVER-LAKE MARGIN COMMUNITY

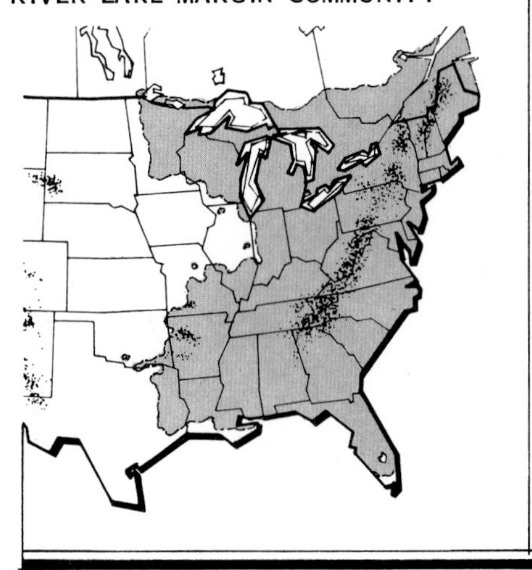

Acer rubrum · Red Maple

Silver Maple grows in bottomlands and along rivers. It is able to withstand flooding for several weeks with no apparent damage.

Silver Maple has big, conspicuous clusters of plump, red flower buds in winter that make the twigs look as though there is something wrong with them. However, when spring arrives, the buds' condition of readiness allows the tree to be the first to bloom.

Both Silver and Red Maple are considered soft maples because of their soft wood. They both grow fast and produce winged samaras in the spring with the leaves, rather than in the fall as the hard maples do. Silver Maple leaves have five, deeply cut lobes. These leaves are silver on the lower surfaces, giving a striking contrast to their green upper surfaces when the wind tosses the branches about.

Silver Maple grows 80 to 90 feet in less than a century, but it is never a strong tree, as wind and ice tear it apart through the years. It is an attractive tree for yards and parks with moist soils. However, its roots are shallow and grass has difficulty competing with Silver Maple because of its high moisture demands.

habitat . . .
FOREST, SAVANNA. lowland wet, lowland wet-mesic, stream, swamp margin and farmstead planting
- zone – 3b

form . . .
GLOBULAR. irregular, large canopy tree (75–100′)
- branching – massive, ASCENDING limbs on stout trunk
- twig – slender, red-brown, smooth, spreading with clustered blunt, rounded buds; bright RED
- bark – silvery-gray, smooth to scaly plates with long, EXFOLIATING ridges

foliage . . .
OPPOSITE. simple, PALMATELY 5-lobed, ovate, dull green leaves (4–6″), DROOPING coarse-toothed margin with reddish petiole; deep, narrow sinuses, silvery-WHITE underside
- color (fall) YELLOW-green
- season – deciduous

flower . . .
CLUSTER. small, red, (3/8–1/2″) dia.; stalkless
- sex – dioecious

fruit . . .
SAMARA. bright red, tan-brown, paired WING (1 1/4″) keys, wide forking clusters; terminal
- season – maturing spring

RIVER-LAKE MARGIN COMMUNITY

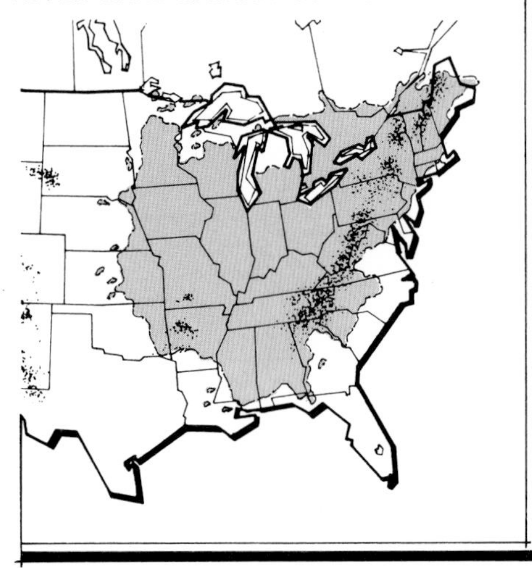

Acer saccharinum · *Silver Maple*

River Birch is a true southern birch that is common to moist lowland depressions, swampy bottomlands, and to isolated bodies of water known as oxbows. It is a fast-growing tree that can reach 30 to 40 feet in twenty years. However, it is not long-lived, maturing in fifty to seventy-five years.

The bark on a young River Birch is paper thin and tinted pinkish-brown. On an old tree that is 50 to 90 feet tall, the bark is thick, deeply grooved, and gray-black.

The leaves are set at an angle from the base. They are sometimes described as diamond-shaped. The flowers of all birches have both male and female catkins. The 3-inch male catkins of River Birch appear in summer but do not flower until the following spring, while the female catkins form small cones that ripen and break apart, scattering their seeds each spring.

River Birch has many trunks and is often used as an ornamental accent tree, displaying much color in its bark; however, it grows fast and soon displays color only in its upper branches.

Betula nigra · River Birch

habitat . . .
FOREST. lowland wet, lowland wet-mesic, bottomlands and stream edge
- zone—5a

form . . .
GLOBULAR. columnar, small canopy tree (50–75′)
- branching—stout, ASCENDING limbs on slightly leaning trunk; often multiple
- twig—glossy red-brown, slender with conspicuous lenticels having woolly (1⁄4″), pointed buds; chestnut brown
- bark—thin, orange, yellow-bronze, PAPERY with exfoliating, horizontal plates

foliage . . .
ALTERNATE. simple, DELTOID-shape, shiny dark green leaves (1½–3″) with double-toothed margin; sharp-pointed tip, whitish beneath; slightly aromatic
- color (fall) YELLOW
- season—deciduous

flower . . .
CATKIN. (male) slender, yellow-green (2–3″), drooping clusters; (female) green, ERECT (1–1½″), upright with reddish tinge
- sex—monoecious

fruit . . .
STROBILE. erect, tan-brown (1½″), ELLIPTIC with 2-winged papery nutlet
- season—maturing summer

RIVER-LAKE MARGIN COMMUNITY

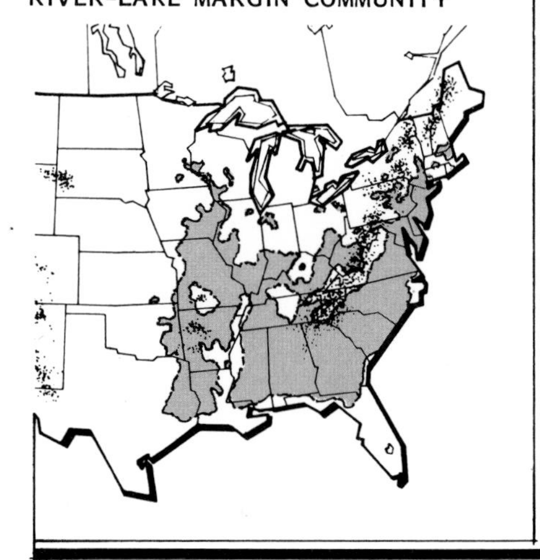

Black Ash is a medium-sized tree, 40 to 70 feet tall, with an open branching pattern. It is a slender tree whose branches reach upward instead of outward like White Ash. The branching is very coarse; even the twigs give the tree an open, branched appearance.

The leaves in autumn have no spectacular color; they turn yellow and then brown, and then they fall all at once on a chilly morning.

Black Ash is primarily a tree of New England, but it also grows in the Midwest, scattered in the low bottomlands along streams and rivers. From its New England home, it has received several common names. It used to be called Hoop Ash or Basket Ash because the wood can be split quite easily into thin, tough splints (Borland, 1983). Baskets produced from Black Ash are among the best to be found.

Black Ash is similar to White Ash, but it has more leaflets in the compound leaf—as many as eleven. Black Ash also has winged samaras that are notched at the ends. This tree grows fast but seldom lives 100 years.

Fraxinus nigra · Black Ash

habitat . . .
FOREST. lowland wet, alluvial flats, lake and stream margin
- zone—2

form . . .
IRREGULAR. narrow, open, small canopy tree (50–75′)
- branching—upright, recurving limbs
- twig—stout, rounded, yellowish brown with conspicuous LENTICLES, having conical-shaped (¼″) buds surrounded by large leaf SCAR; dark brown to BLACK
- bark—smooth, light gray, scaly, fissured with loose, CORKY ridges

foliage . . .
OPPOSITE. pinnately COMPOUND (7–11) leaflets, oblong-lanceolate, dark green leaves (3–5″) with fine-toothed margin; stalkless, pale beneath
- color (fall) YELLOW-green
- season—deciduous

flower . . .
CLUSTER. small, deep PURPLISH black (⅛″) compact clusters
- sex—dioecious

fruit . . .
SAMARA. tan-brown, single KEY (1–1½″) dense, drooping clusters; wing oblong
- season—maturing late summer

RIVER-LAKE MARGIN COMMUNITY

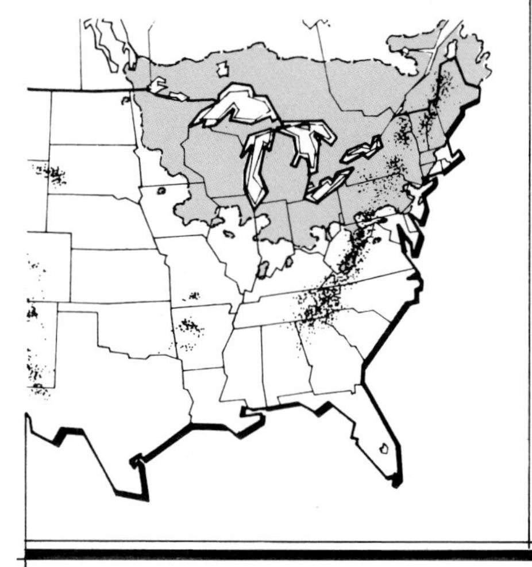

Black Tupelo has several names: Black Gum, Sour Gum, and Pepperidge. This last name, Pepperidge, was the English name for barberries because, "like the barberries, the Tupelo berries are acid" (Collingwood and Brush, 1974). This name later became the inspiration for an American food, pepperidge bread.

Black Tupelo is a tree that is often found in moist, rich bottomland soils, and it is recommended for landscaping ponds. It is quite adaptable and often grows in areas very different from its native habitat.

In its early years, Black Tupelo has a very symmetrical, pyramidal shape with all but the top branches being virtually horizontal. One very prominent characteristic of the species is the rigidity and zig-zaggedness of its branches, which could be due to its tough cross-grained wood.

Another attribute of Black Tupelo is its thick, glossy, deep green leaves. In autumn, these leaves turn to a deep red glow that only sumacs can come close to equaling. The fruit ripens to black at the same time the leaves are turning red. Though the fruit is sour, it provides a feast for many species of birds.

Black Tupelo grows at a moderate rate and has a medium life span.

habitat . . .
FOREST. upland dry, lowland wet, lowland wet-mesic and alluvial flats
- zone – 5a

form . . .
CONICAL. irregular, small canopy tree (50–75′)
- branching – slender, CROOKED horizontal limbs
- twig – slender, green-reddish brown, short-SPURRED with flaky skin having OVOID (1/4″), yellow-brown buds; hairy
- bark – thick, black to REDDISH brown, deeply fissured with conspicuous rectangular, BLOCKY ridges

foliage . . .
ALTERNATE. simple, oblong-obovate, dark SHINY green leaves (2–5″) with entire to wavy margin; LEATHERY, clustered near twig ends
- color (fall) SCARLET-red
- season – deciduous

flower . . .
CLUSTER. small, yellow-greenish white, (3/16–1/4″) dia.; long-stalked at leaf axils
- sex – dioecious

fruit . . .
BERRY. small, blue-BLACK (3/8–1/2″), elliptical clusters 1–3 along stem
- season – maturing late autumn

RIVER-LAKE MARGIN COMMUNITY

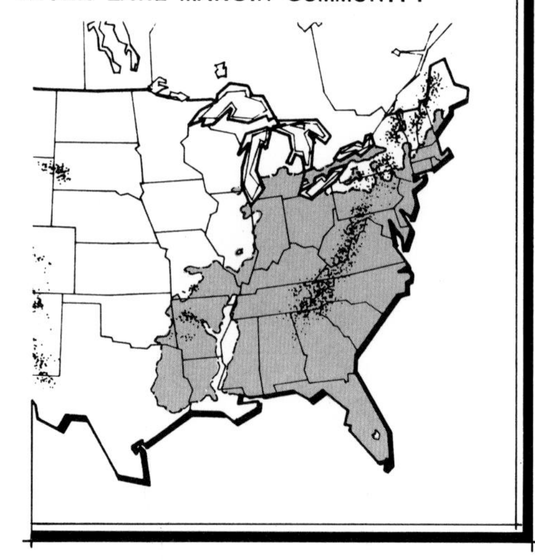

Nyssa sylvatica · Black Tupelo

American Planetree, or Sycamore, is an open, globular tree that is common along streams and rivers and in rich, bottomland soils. It is a very fast grower; 70 feet in 20 years is not uncommon. Many trees live for 350 years.

Sycamore's main attribute is its bark. At the base of the trunk, the bark is dark brown and rather deeply fissured; on the limbs and among the upper branches, it is a mottled mixture of creamy white blotches and darker, older areas where the characteristic peeling process has not yet begun.

In summer, the huge lobed leaves mask the bark coloration; in autumn and winter, the tree's startling, unique bark is once again revealed. The leaves are maple-shaped and entirely cover the bud at the leaf base. After the leaves fall, a multitude of ball-shaped seed clusters are revealed, hanging from twig tips as though the tree had been decorated for Christmas.

Sycamore is a good choice where a really large and easily transplanted tree is needed. The tree stands pruning well and is resistant to wind and storm damage. If planted with other Sycamores, the likelihood of having an anthracnose problem is greatly increased. The disease slows the development of leaves in early summer, causing the browning of leaf edges, but it does no damage to the tree.

habitat . . .
FOREST. lowland wet, lowland wet-mesic, stream and lake margin
- zone – 4a

form . . .
GLOBULAR. open, large canopy tree (75–100′)
- branching – massive, PICTURESQUE limbs on straight trunk
- twig – stout, orange-brown, green, slender ZIG-ZAG with DOME-shaped, red-brown bud, encircled by leaf scar
- bark – exfoliating, creamy white, MOTTLED patches tan-brown; base with squarish plates (old)

foliage . . .
ALTERNATE. simple, large, PALMATELY (3–5) lobed DELTOID-shape bright green leaves (4–8″) with coarse-toothed margin; veins hairy beneath
- color (fall) tan-BROWN
- season – deciduous

flower . . .
CLUSTER. (male) yellow, small; (female) red, (3/8–1/2″) dia., BALL-like clusters on long, slender stems; pendulous
- sex – monoecious

fruit . . .
MULTIPLE. tan-brown, globular, (1″) dia. on LONG narrow stalk; nutlets with HAIRY tufts
- season – maturing in autumn, PERSISTING through winter

RIVER-LAKE MARGIN COMMUNITY

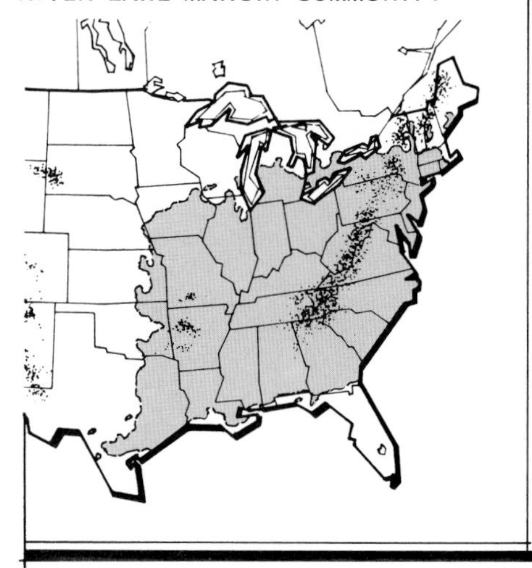

Platanus occidentalis · American Planetree

Eastern Poplar, or Cottonwood, is a stately tree that usually towers above all others on the floodplain, growing 100 or more feet tall. It has an extremely wide range, from Quebec to Florida and westward to the Rocky Mountains. It may survive for more than 125 years in any of these locations.

Cottonwood usually grows best in moist areas near rivers and streams, but it tolerates a variety of sites, being one of only a few trees that can withstand even difficult city conditions.

Cottonwood has a reputation of being a brittle tree, for its wood is light and soft. Though it suffers considerable branch breakage in windy locations, damage to the tree is generally minor and broken branches are replaced surprisingly quickly.

The greatest objection to Cottonwood comes from the fact that the female trees produce clouds of cottony seeds each spring that clutter lawns, garages, and screen doors. This is a temporary condition that can be tolerated when one compares this inconvenience with the value of Cottonwood's shade and with the soothing wind sounds from the rattling leaves of one of these gentle giants.

habitat . . .
FOREST, SAVANNA. lowland wet, lowland wet-mesic, floodplain and stream edge
- zone—3b

form . . .
OVOID. globular, large canopy tree (75–100′)
- branching—high, ARCHING limbs with descending tips on massive trunk
- twig—stout, creamy yellow-brown, smooth with sharp-POINTED (3/4″), shiny brown clustered buds; RESINOUS
- bark—smooth, yellowish-green (young) becoming thick, gray-brown with MASSIVE, deep upright furrows

foliage . . .
ALTERNATE. simple, broad, DELTOID-shaped, bright green leaves (3–7″) with coarse, in-curved toothed margin; glossy on long, flat petiole
- color (fall) pale YELLOW-orange
- season—deciduous

flower . . .
CATKIN. linear, (male) RED, (female) yellow-green (2–4″), drooping clusters
- sex—dioecious

fruit . . .
CAPSULE. small, yellow-green (3/8″), drooping, CATKIN-like (4–8″) clusters with COTTONY seeds
- season—maturing early summer

RIVER-LAKE MARGIN COMMUNITY

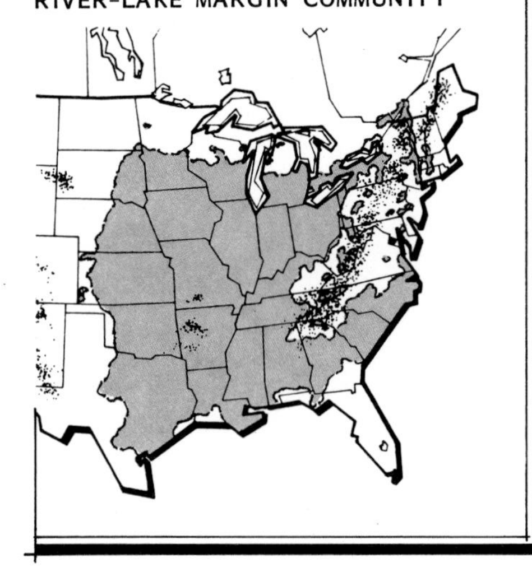

Populus deltoides · *Eastern Poplar*

Swamp White Oak is a large canopy tree that is common in the moist, deep soils of stream borders, lake margins, and alluvial flats (areas separated by distinct slope from the floodplain or water course). It grows especially well in southern states and can reach heights of 75 to 100 feet with a trunk 8 feet in diameter. It is a good lumber tree and is sold commercially as White Oak.

The leaves of Swamp White Oak are typical of the White Oak Family because the undersides are silvery, in contrast with the warm green of the upper surfaces. However, the shallow lobes or scalloped edges of the leaves seem quite different from oaks whose leaves have deeply cut sinuses.

Swamp White Oak differs from other oaks in its tendency to have exfoliating bark on the secondary branches. It is also one of the few oaks that supports its acorns on long stems. Its fall leaf color, although not spectacular, is a respectable golden brown.

Swamp White Oak is one of the fastest growing oaks, adding 1½ to 2 feet a year. It generally matures in 125 years. It does require acid soils and, like Pin Oak, is severely affected when it gets inadequate amounts of iron.

habitat . . .
FOREST. lowland wet-mesic, lowland wet, alluvial flats, lake and stream edge
- zone – 4a

form . . .
OVOID. open, large canopy tree (75–100′)
- branching – horizontal, UPRIGHT limbs
- twig – stout, red-brown, slender with RAISED lenticels having clustered chestnut-brown, GLOBULAR buds; blunt ends
- bark – gray-brown, deep-furrowed with blocky to scaly ridges; PAPERY, exfoliating (young)

foliage . . .
ALTERNATE. simple, shallow, (5–10) LOBED obovate-oblong, dark green leaves (5–7″) with wavy margin; WEDGE-shaped base, white, hairy beneath to shiny above
- color (fall) YELLOW-brown
- season – deciduous

flower . . .
CATKIN. (male) yellow-green (2–3″), drooping clusters
- sex – monoecious

fruit . . .
ACORN. long-stemmed, tan-brown (¾–1¼″), PAIRED clusters; ⅓ enclosed by slightly fringed, scaly cup
- season – maturing first year

RIVER-LAKE MARGIN COMMUNITY

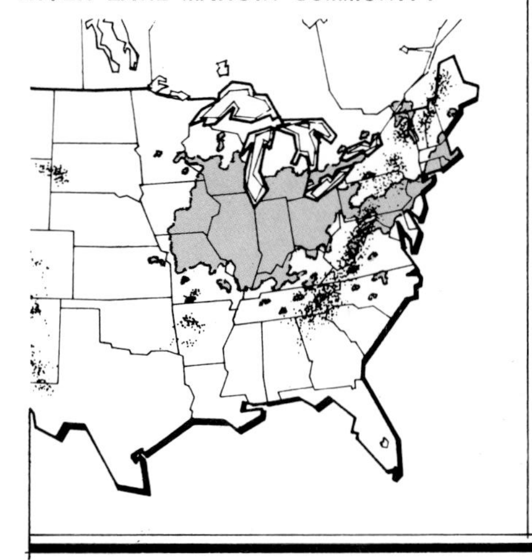

Quercus bicolor · *Swamp White Oak*

Pin Oak is a notable exception to the rugged appearance of other oaks. Instead of the gnarled, massive qualities so common to most oaks, Pin Oak suggests refinement. The leaves are finely chiseled with graceful lines. The branches are slender, horizontal, and drooping. The silhouette is symmetrical. Even the acorns are small and button-like with tight, scaled caps.

Pin Oak also has an unusual tolerance for bottomland soils and poorly drained clay flats. Sometimes it is found growing in swamps, having no apparent problem with wet feet.

Most oaks are considered slow growing and rather difficult to transplant. However, Pin Oak grows as rapidly as many other desirable shade trees, and its fibrous root system allows easy transplanting.

The leaves of most oaks are held long into the fall, but Pin Oak holds many of its leaves into early spring. The foliage is one of Pin Oak's best qualities, being smooth and shiny and having a rich maroon fall color. In some areas where the soil is lacking iron, this deficiency can cause yellow coloration in the leaves through the summer months; if not treated, this problem can eventually kill the tree.

Pin Oak is a beautiful tree in a group or as a single specimen, but it needs room to spread and to allow its lower limbs to droop and grace the ground below.

Quercus palustris · Pin Oak

habitat . . .
FOREST. lowland wet-mesic, lowland wet and alluvial flats
- zone—5a

form . . .
CONICAL. dense, small canopy tree (50–75′)
- branching—HORIZONTAL, descending limbs, often SWEEPING to the ground
- twig—smooth, green to red-brown, slender, PIN-like with sharp-pointed, SHINY clustered buds; red-brown
- bark—smooth, ASH-GRAY to black, shallow fissured with scaly ridges on upper trunk

foliage . . .
ALTERNATE. simple, pinnately (5–7) LOBED, obovate shiny dark green leaves (3–7″) with irregular, WIDE, deep sinuses; bristle-tipped
- color (fall) SCARLET-red
- season—deciduous

flower . . .
CATKIN. (male) yellow-green (2–3″), slender, drooping clusters; (female) in leaf axils
- sex—monoecious

fruit . . .
ACORN. red-brown (½″) flattened, DOME-shaped; ¼ enclosed by shallow, SAUCER-like cup
- season—maturing second year

RIVER-LAKE MARGIN COMMUNITY

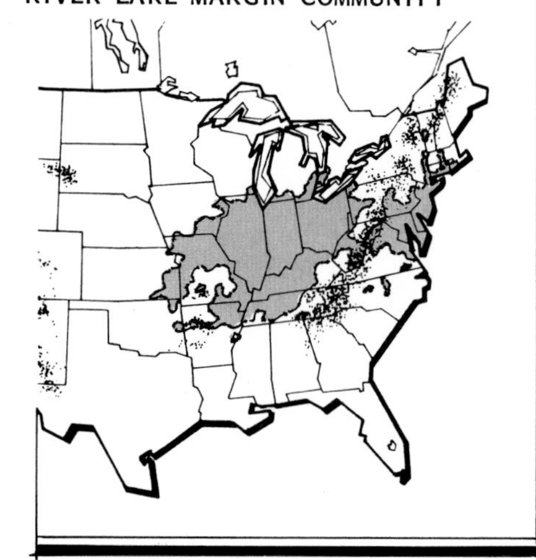

Black Willow is a shrub or tree, 10 to 60 feet tall, that is common to streams and marshes. It lives in wet ground where few other plants can survive.

The twigs are an important feature of Black Willow. They are conspicuous because of their polished yellow that in winter catches the sun's rays to create a dramatic effect against the blue sky. The twigs are also important for propagation: all it takes to start a new Black Willow is to thrust a twig into moist ground. Black Willow trees cast off twigs the way other trees cast off fruit, thus providing a way for new plants to get started.

The bark of an old Black Willow is heavily ridged and often swirled and distorted. The wood is soft and almost useless for timber. However, before corrugated cardboard came into use, grocery items were crated in boxes made from Black Willow.

Male and female blossoms are found on separate trees. The female, or pistillate, catkins produce tiny, pear-shaped pods containing seeds with silky down attached to them. The leaves of Black Willow are long and narrow, 1½ inches wide and 5 inches long. They taper sharply and have small teeth along each edge like a fine saw blade.

Black Willow is a slender tree with an open, attractive crown. Because of limb breakage in storms, it should not to be planted near buildings.

Salix nigra · Black Willow

habitat . . .
SAVANNA. lowland wet, lowland wet-mesic, swamps and stream edge
- zone – 3a

form . . .
COLUMNAR. irregular, large understory tree (35–50′)
- branching – ascending limbs often on MULTIPLE, leaning trunks
- twig – smooth, orange, yellow-green, slender, glossy with incurved, one-scale yellow-brown buds; HUGGING twig
- bark – fibrous, gray-brown, DEEP-furrowed with black scaly ridges

foliage . . .
ALTERNATE. simple, oblong-lanceolate, NARROW, bright green leaves (3–5″) with fine-serrated margin; slightly long-pointed, curved to one side, short-stalked
- color (fall) YELLOW-green
- season – deciduous

flower . . .
CATKIN. slender, yellow-green (1–3″); elliptic with scales hairy
- sex – dioecious

fruit . . .
STROBILE. small, yellow-green (3/16″), CATKIN-like (1–3″) cluster releasing seeds with SILKY white hairs; tufted
- season – maturing early summer

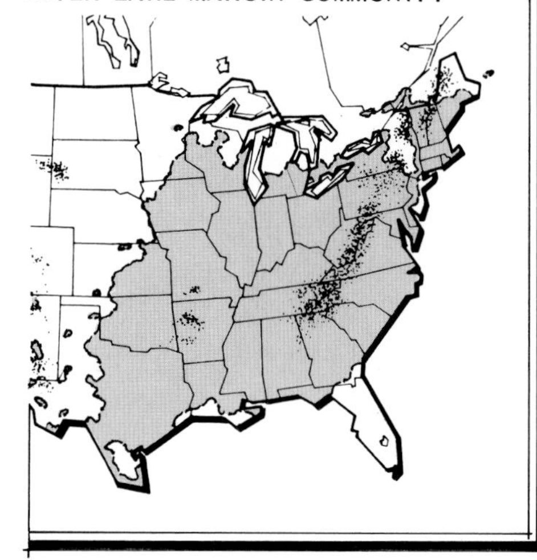

Common Baldcypress is more difficult to approach in native surroundings than to identify. It grows in southern swamps and moist, alluvial bottomlands where it commonly has a "flared or buttressed trunk rising out of mud or shallow water" (Kieran, 1954). Surrounding the tree are "knees," cone-shaped roots that stick up above the water surface. There is much discussion about whether the purpose of these knees is to anchor the tree or to provide an air supply for roots that are constantly under water.

When the tree is planted among strangers in northern gardens, it puts on its best manners. There is not a hint of hollow buttresses or bare knees. In fact, in its early years, Baldcypress grows beautifully and in perfect symmetry, with a feathery lightness in its leafy spray.

Baldcypress has needles for foliage and produces cones, but it is not an evergreen. Like Tamarack, or Larch, it is an exception and drops its needles each fall. The needles do not fall individually, but the feathery branchlets drop intact from the tree, for each one is a single leaf, although a compound one.

The fruit is a round cone about 1 inch in diameter. The bark is a silvery cinnamon-red. Baldcypress wood is light but durable; it has been used for casks, shingles, and railroad ties.

Common Baldcypress tolerates many soils and the dry conditions of the North. It has become a popular ornamental because of its rusty red fall color and conical shape. It grows at a medium to fast rate and lives 400 to 600 years.

Taxodium distichum · Common Baldcypress

habitat . . .
FOREST. lowland wet-mesic, lowland wet, alluvial flats, swamp and river edge
- zone—5a

form . . .
COLUMNAR. conical, large canopy tree (75–110′)
- branching—horizontal to ascending limbs
- twig—stout, green to tan-brown, short with small (inconspicuous) buds; brown
- bark—smooth, red-brown, scaly to fibrous, PEELING strips; base deeply ridged often with KNEES, arising from roots on wetter sites

foliage . . .
LINEAR. needle-like, pinnately COMPOUND, 2-ranked, blue-green leaves (3/8–3/4″), flat with FEATHER-like appearance; alternate in spirals
- color (fall) maroon-PURPLE to brown
- season—deciduous

flower . . .
CATKIN. (male) yellow-green (4–12″), drooping, slender clusters; (female) green, small (1/8″), cone-like
- sex—monoecious

fruit . . .
CONE. globular, PURPLISH brown, (1″) dia. with wrinkled scales, 4-ANGLED; pendant, often 1 to 2 at end of twigs
- season—maturing late autumn

RIVER–LAKE MARGIN COMMUNITY

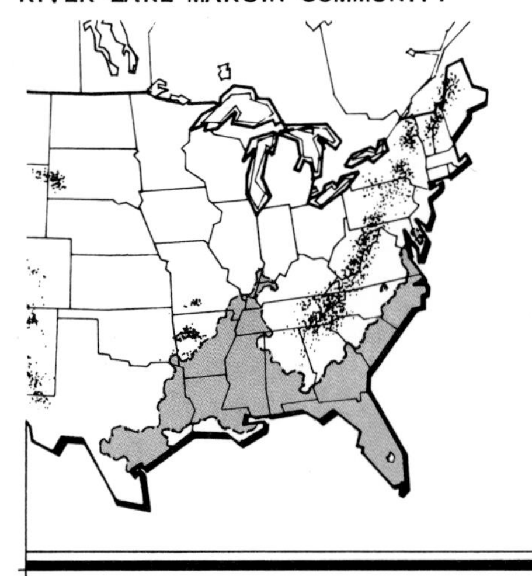

Eastern Arborvitae, or White Cedar, is a tree of the Northern Forest that grows along lakes and in marshes. Like so many other wetland species, it also grows well on dry ground.

In its crowded, water-soaked native environment, White Cedar is a slender, not well-branched tree. In the open where it is free from competition, White Cedar improves in form and density. In its northern range, it seldom reaches a height of more than 40 feet; in the South, it grows to twice that size.

The wood is slightly fragrant and extremely resistant to decay. Not long ago, in low-lying parts of the Atlantic coast, there were still in operation what might be called White Cedar "mines." In these mines, or swamps of buried trunks, great trees from many centuries ago were excavated and converted into shingles, cabinet materials, and planking for small high-grade boats (Borland, 1983).

White Cedar is often used as a specimen plant. It also produces an excellent hedge that is quite dense. White Cedar serves as well as any other plant for sound barrier plantings. Its most memorable attribute, in my thinking, is the strong, pleasant, aromatic scent given off by the foliage when crushed. That scent immediately transports me to the pristine beauty of the Canadian Boundary Waters.

Eastern Arborvitae trees are attractive to birds for cover. Its seeds are a preferred food source of Pine Siskins (Degraaf and Whitman, 1979).

Thuja occidentalis · Eastern Arborvitae

habitat . . .
FOREST. upland mesic, upland mesic-dry, lowland wet, lowland wet-mesic, lake margin and open rocky slopes
- zone – 2

form . . .
COLUMNAR. conical, small canopy tree (50–75′)
- branching – horizontal, SPREADING limbs with ascending tips; singular to MULTIPLE trunk
- twig – slender, yellow-green in flattened SPRAYS with (inconspicuous) naked buds
- bark – fibrous, red-brown, exfoliating to PEELING vertical strips

foliage . . .
SCALE-like. small, yellow-green (1/8″), rounded lateral scales; flat, FAN-shaped with glandular, conspicuous dots
- color – YELLOWISH BROWN-green
- season – evergreen

flower . . .
CONE. small, red-brown to YELLOW, 1/16–1/8″ across at scale ends
- sex – monoecious

fruit . . .
CONE. erect, tan-brown (1/8–1/2″), elliptic, EGG-shaped with (8–10) scales; leathery
- season – maturing following autumn

RIVER-LAKE MARGIN COMMUNITY

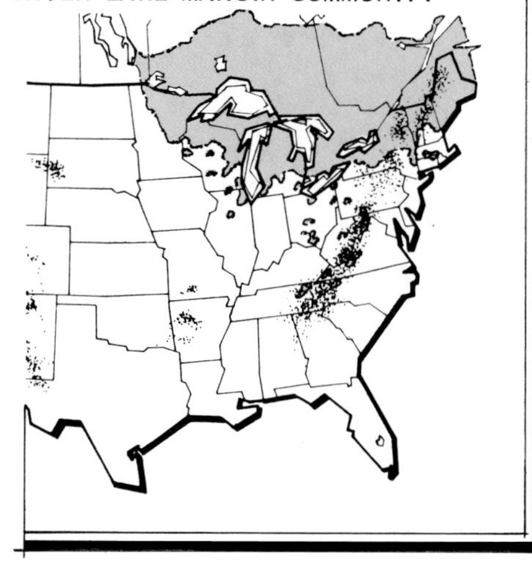

Indigobush Amorpha, or False Indigo, is a shrub that grows 5 to 20 feet tall. It prefers the edges of streams and ponds, and it often grows in dense thickets. Indigobush Amorpha is in the Legume Family and has compound leaves that are 6 to 16 inches long. The small fruit pods, which never open, usually have two seeds and are high in nitrogen.

The word "amorpha" means formless or deformed. The flower of Indigobush Amorpha consists of only one petal. This petal, or banner, wraps itself around the stamens and pistil, trying to do the work of several petals in protecting the central parts of the flower. This petal is a deep, rich purple, and the protruding anthers are brilliant orange. When the sun highlights these unusual flower clusters, they glow luridly.

Indigobush Amorpha flowers are pea-like and are clustered in dense, terminal racemes. The flowers at the base of the spike bloom first, and the flowering proceeds spirally upward, prolonging the blooming period for a considerable time (Keeler, 1903).

The shrub can be used effectively as an ornamental in moist sites. After the blooming period is over, Indigobush Amorpha looks very much like a Black Locust seedling.

Amorpha fruticosa · Indigobush Amorpha

habitat . . .
SAVANNA. lowland wet, lowland mesic-wet, alluvial flats, lake and stream edge
- zone – 4

form . . .
GLOBULAR. large, upright shrub (4–12′)
- branching – open, stiff to arching limbs
- twig – slender, gray to tan-brown, rounded or finely GROOVED with smooth gray-brown buds; bundle scars (3) often appearing above each other
- bark – smooth, tan-brown, with WARTY appearance

foliage . . .
ALTERNATE. odd-pinnately COMPOUND, (11–25) leaflets oblong-elliptic, dull yellow-green leaves (4–12″) with entire margin; thick, base somewhat rounded
- color (fall) YELLOW-green
- season – deciduous

flower . . .
SPIKE. erect, PURPLISH blue (3–6″), dense clusters with PEA-like (3/8″) bloom; yellow-orange anthers
- sex – monoecious

fruit . . .
LEGUME. brown (1/4″), compressed, BANANA-like pod; terminal spike
- season – maturing late summer

RIVER-LAKE MARGIN COMMUNITY

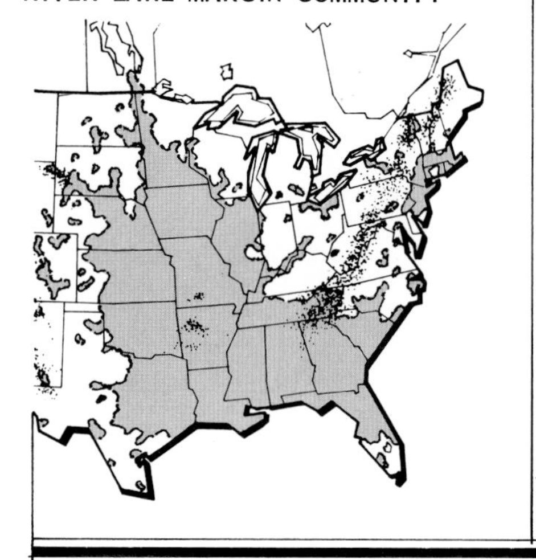

Common Buttonbush, sometimes called Honey Balls or Globeflower, is a tall shrub that grows near lakes. It sometimes grows in shallow water and can form dense stands or thickets.

The flower of Buttonbush is the plant's most distinctive attraction. It is not unusual in nature to find round fruit forms, but round flowers are unusual. Yet the flower cluster of Buttonbush is a perfect globe. Blooming in late July and August, the flowers provide an abundance of nectar. The fragrant flower, so loved by bees, usually hangs over water and must be easier to pollinate than to photograph.

The smaller ball-like heads of dry fruit that follow mature in September. The seeds, a food source throughout winter, are often eaten by flocks of waterfowl. The large leaves, 3 to 6 inches long, are opposite and are found in whorls of three at nodes along the stems.

In American gardens, Common Buttonbush is seldom seen; in Europe, it is planted for its unusual flower and its late season of bloom.

habitat . . .
SAVANNA. lowland wet, lowland wet-mesic, stream and water edge
- zone – 4

form . . .
IRREGULAR. globular, large shrub (3–10′)
- branching – stiff, spreading limbs; often rambling
- twig – stiff, tan-brown with small (suppressed) buds
- bark – tan-brown with PEELING vertical strips

foliage . . .
OPPOSITE. simple, WHORLS of 3, elliptic, ovate-lanceolate, dark green leaves (3–6″) with entire margin; both ends tapering
- color (fall) yellowish BROWN to green
- season – deciduous

flower . . .
GLOBULAR. creamy WHITE (1–1½″), dense balls with clustered (⅜″), PIN-like stypes; appearing in leaf axils or along stalks
- sex – monoecious

fruit . . .
MULTIPLE. dense, reddish green (⅜–½″), BALL-like cluster
- season – maturing late summer

RIVER-LAKE MARGIN COMMUNITY

Cephalanthus occidentalis · Common Buttonbush

Silky Dogwood is a wide, spreading shrub that grows 8 to 10 feet tall. It is common in the North and is found on stream banks, in wet meadows, and along fencerow ditches.

It is the latest blooming plant in the Dogwood Family to flower, usually coming into bloom about the third week of June. The flower, much like other shrub dogwoods, is a small, creamy white cluster, but the fruit is quite different. Instead of the usual white berries, Silky Dogwood produces pale blue clusters of bitter but aromatic berries that are high in food value to wildlife.

Indians used the plant in kinnikinnik, a favorite smoking mixture consisting of tobacco and wood scrapings from Silky Dogwood (Grimm, 1957).

Silky Dogwood is sometimes mistaken for Redosier Dogwood. Both grow in shady places, but Silky Dogwood has twigs that are first grayish-green and later become purplish red. Redosier Dogwood, on the other hand, has brilliant red stems and white fruit.

Silky Dogwood is an excellent plant for a shrub border and for use in preventing lake or stream bank erosion. Since the shrub grows quite well under cultivation on well-drained sites, it is recommended by the U.S. Soil Conservation Service for both erosion control and wildlife habitat plantings (Grimm, 1957).

habitat . . .
FOREST, SAVANNA. lowland wet, lowland wet-mesic, stream and water edge
- zone – 4

form . . .
GLOBULAR. upright, large shrub (6–10′)
- branching – compact, spreading limbs
- twig – glossy, reddish brown, HAIRY with tan-brown pith having VELVETY gray, narrow, blunt buds
- bark – smooth, gray-green to reddish brown

foliage . . .
OPPOSITE. simple, elliptic, ovate-lanceolate, dark green leaves (2–4″) with entire margin; veins PARALLEL to leaf edge with reddish hairs in veins beneath
- color (fall) red-MAROON
- season – deciduous

flower . . .
CLUSTER. yellowish white (2½–3″), FLAT-TOPPED clusters with dense (¼″) blooms
- sex – monoecious

fruit . . .
BERRY. small, blue (¼″), GLOBULAR clusters to (3″) across
- season – maturing late summer

RIVER-LAKE MARGIN COMMUNITY

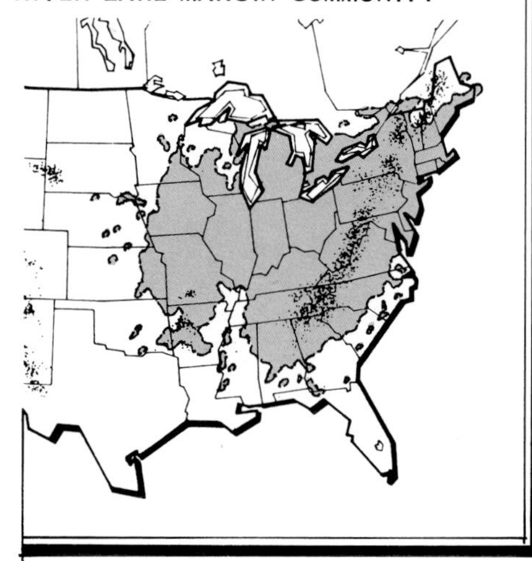

Cornus amomum · Silky Dogwood

Redosier Dogwood grows in thickets in dry soils, but it is most commonly found in wet sites along stream banks or lakes. Redosier Dogwood grows rapidly in any good soil and is an excellent plant for mass plantings and hedge borders. It is a shrub tolerant of some shade. A single plant, spreading by means of underground shoots, can transform itself into a thicket.

In winter, the brilliant red stems are attractive against a snowy background. Redosier Dogwood is also good in a shrub border and contrasts well with an evergreen background. Its fruit is eaten by many birds. Its upright twigs are preferred as nesting sites by goldfinches (Degraaf and Whitman, 1979).

The flowers of Redosier Dogwood are much like many other shrub dogwoods: flat clusters of yellowish-white flowers that mature in the fall to a cluster of white berries about the size of small peas, each on a red stem. The berries seem unpalatable, but Indians used to eat them. They also made bows from the wood and brewed a concoction of leaves and bark to treat colds and fevers.

habitat . . .
FOREST, SAVANNA. upland mesic-dry, lowland wet-mesic, stream and water edge
- zone—2

form . . .
GLOBULAR. spreading, large shrub (6–12′)
- branching—DENSE, ascending limbs
- twig—dark blood RED, erect stems with VELVETY gray-brown, long-pointed buds; base swollen
- bark—smooth, bright red prostrate stems with large, white PITH

foliage . . .
OPPOSITE. simple, elliptic-ovate, glossy green leaves (2–4″) with entire margin; base WEDGE-shaped veins (5–6) parallel to leaf edge, tapering apex
- color (fall) red-MAROON
- season—deciduous

flower . . .
CLUSTER. white (2″) FLAT-TOPPED cluster with dense (1/4″) blooms; petals (4)
- sex—monoecious

fruit . . .
BERRY. small, white turning BLUE (1/4″), globular clusters (3½″) across
- season—maturing late summer

RIVER-LAKE MARGIN COMMUNITY

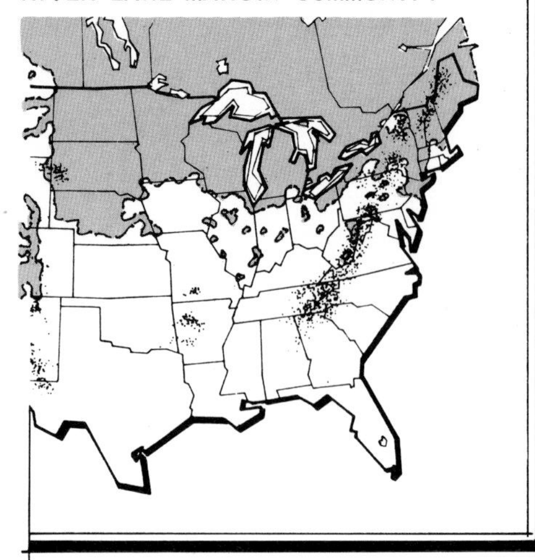

Cornus stolonifera · Redosier Dogwood

Common Winterberry has an unrivaled winter color provided by the persistent berries of the native holly. Still, this shrub is virtually unknown.

Its fruit, which ripens in September or October, is frequently produced in great abundance. The berries remain on the branches long after the leaves have fallen and often remain throughout the winter. They are eaten by Ruffed Grouse, Cedar Waxwings, and other birds, and they are used in winter bouquets (Grimm, 1957).

The leaves of Common Winterberry are not evergreen, like other hollies, nor do they have sharp teeth, or spines, on leaf margins. The flowers of this shrub are small, yellow-green, and inconspicuous. The plant becomes noticeable after the leaves have fallen, revealing the deep red berries.

Common Winterberry is sometimes referred to as Black Alder, because it frequents the same haunts. It tolerates full sun but prefers partial or deep shade in moist soil near ponds, lakes, or marshes.

Common Winterberry should be used more as an ornamental shrub in parks and gardens because it equals or surpasses any imported plant that is known for its winter color in both brilliancy and beauty.

habitat . . .
FOREST, SAVANNA. upland mesic-dry, lowland wet, lowland wet-mesic, pond and stream edge
- zone – 3

form . . .
GLOBULAR. upright, medium shrub (3–10′)
- branching – ERECT, spreading limbs
- twig – stout, tan-brown with brown, blunt buds; broad-pointed scales
- bark – smooth, tan-grayish green

foliage . . .
ALTERNATE. simple, elliptic, oblong-lanceolate, LEATHERY bright green leaves (1½–4″) with coarse-toothed margin; dull above, hairy veins beneath
- color (fall) GREEN
- season – deciduous

flower . . .
CLUSTER. small, yellowish white (⅛–¼″), dense clusters at leaf axils
- sex – dioecious

fruit . . .
BERRY. small, RED (¼″), dense clusters along stem; turning black
- season – maturing late autumn, PERSISTING

RIVER-LAKE MARGIN COMMUNITY

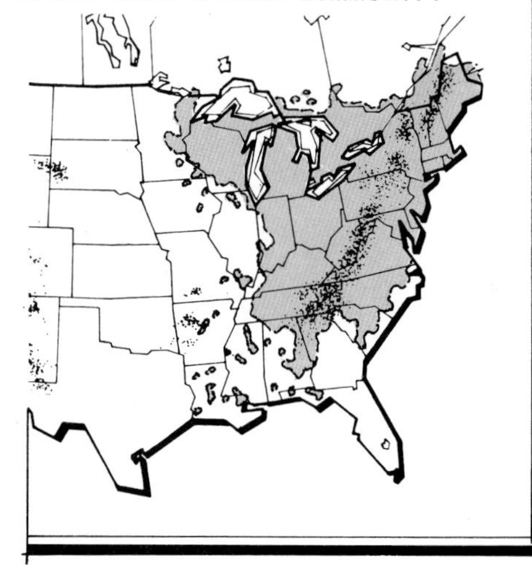

Ilex verticillata · Common Winterberry

Northern Bayberry is a very adaptable shrub. It grows in marshes and on sand dunes. The leaves turn brown in the fall but tend to remain throughout winter if protected from winter winds. When the leaves do fall, clusters of waxy, gray berries are revealed along the stems.

The berries of this shrub are valued by birds for food and by people for their appearance and for use in making candles. Bayberry wax is obtained by boiling the berries in water. The wax dissolves, rises to the surface, and hardens when cooled. It is estimated that about one-third of the berry consists of wax (Keeler, 1903). The leaves are used by some as a substitute for bay leaves to season meats and soups.

Northern Bayberry is an attractive plant that is becoming more popular as an ornamental for use in a shrub border. It has been recommended by the U.S. Soil Conservation Service for use in erosion control and wildlife plantings (Grimm, 1957).

habitat . . .
FOREST, SAVANNA. upland dry, upland mesic-dry, steep rocky slopes, sandy soils, stream banks and lake shores
- zone – 2

form . . .
UPRIGHT. rounded, medium shrub (3–8′)
- branching – SYMMETRICAL, spreading upright limbs
- twig – stout, gray PUBESCENT, stiff with whitish, globular end buds
- bark – smooth, gray

foliage . . .
ALTERNATE. simple, oblong-obovate, bright green leaves (1–5″) with apex toothed or entire; somewhat hairy with RESINOUS dots
- color (fall) MAROON-tan
- season – broadleaf evergreen; spicy AROMATIC fragrance

flower . . .
CATKIN. small, yellow-green (½–¾″) clusters along stem
- sex – dioecious or monoecious

fruit . . .
BERRY. small, grayish WHITE (⅛″), globular, dense clusters at leaf axils; WAXY appearance
- season – maturing late summer, PERSISTING through winter

RIVER-LAKE MARGIN COMMUNITY

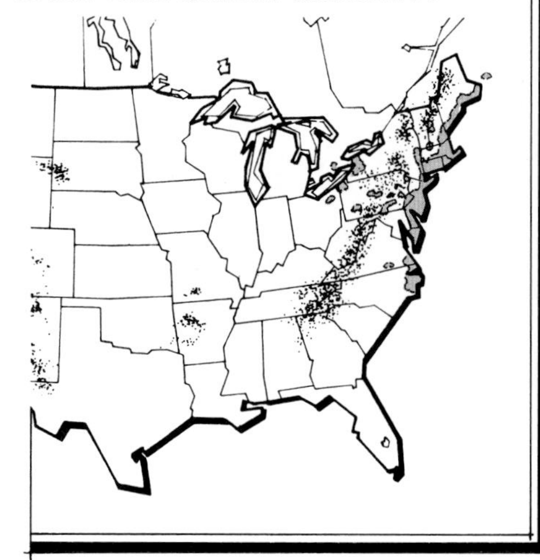

Myrica pensylvanica · Northern Bayberry

Common Ninebark is easily mistaken for Vanhoutte Spirea when it is in bloom. However, Ninebark has much more to offer than its spirea-like flowers. After the petals fall, the bladder fruit begins to form, attracting attention because the weight of the fruit arches the slender branches almost to the ground. The inflated pod clusters change from green to russet, finally becoming rose-purple. These pods are held throughout winter and move with the wind, rattling and scattering their seeds.

Ninebark's golden yellow fall leaf color is also attractive.

In the winter when the leaves have fallen, the bark is revealed, showing the thin papery layers that exfoliate from the main stems. This last characteristic is where the plant gets the name Ninebark.

Common Ninebark is found along river banks and moist, rocky slopes, but it grows well in any good garden soil. It is an excellent plant for a shrub border and provides a good companion to Gray Dogwood and Smooth Sumac for fall color variation.

habitat . . .
FOREST, SAVANNA. upland dry, upland mesic-dry, steep rocky slopes and rock outcrops
- zone – 2

form . . .
OBOVOID. upright – medium large shrub (3–10′)
- branching – ARCHING to recurving limbs
- twig – stout, tan-silvery gray; GLABROUS with small, brown buds; many scaled
- bark – papery, tan-brown, PEELING in long thin strips

foliage . . .
ALTERNATE. simple, 3-LOBED ovate, bright yellow-green leaves (1–5″) with irregular, round-toothed margin; leaf veins, tomentose beneath
- color (fall) golden YELLOW
- season – deciduous

flower . . .
CLUSTER. small, pinkish WHITE (¼–⅜″) dense, UMBRELLA-like clusters (1–2″) across; terminal
- sex – monoecious

fruit . . .
POD. reddish brown, glabrous, dense clusters; (3–5) parted
- season – maturing late autumn, PERSISTING

RIVER-LAKE MARGIN COMMUNITY

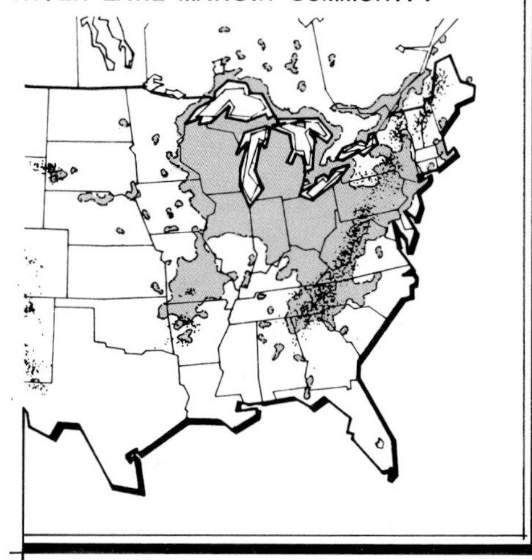

Physocarpus opulifolius · Common Ninebark

Sandbar Willow is a shrub or small tree that often grows less than 20 feet tall. It is a common plant on river sandbars. It grows there in thickets until the land becomes stable enough to support other species, such as Silver Maple and Cottonwood.

Sandbar is an appropriate name for this willow, but it has another common name, Longleaf Willow. In general, willows are characterized by their simple leaves that are much longer than they are wide. But Sandbar Willow is the narrowest of them. The leaves are lance-shaped, 2 to 5 inches long, and ¼ to ½ inch wide. The teeth along the margin of this willow leaf are widely spaced, making the plant easy to identify.

The soft yellow bloom of the catkins appears in April and May. The fruit matures in June or July, producing many whitish, cottony seeds. These seeds can germinate in seven days after they drop to moist ground—a prospect that seems very likely on sandbars along rivers and streams.

Sandbar Willow is not commonly used in the home landscape but could be planted in solid thickets to produce wildlife habitat and for holding the soil on low-lying erodible sites.

Salix interior · *Sandbar Willow*

habitat . . .
SAVANNA. lowland wet, lowland wet-mesic, sandbars, silty alluvial bottomlands, streams and pond edge
- zone–2b

form . . .
OBOVOID. leggy, small understory tree (20–35′)
- branching—erect limbs often forming DENSE thickets
- twig—slender, yellowish brown, glabrous with small, rounded buds; convex
- bark—gray-brown, SLIGHTLY furrowed with scaly ridges

foliage . . .
ALTERNATE. simple, NARROW, lanceolate, bright green leaves (3–6″) with wide marginal teeth; stipules conspicuous, glossy with yellowish midrib
- color (fall) YELLOW-green
- season—deciduous

flower . . .
CATKIN. slender, yellow-green (1–2″), elliptic clusters of 2–5 in axils
- sex—dioecious

fruit . . .
STROBILE. small, yellow-green (3/16″), CATKIN-like (1–4″) cluster; sessile, releasing seeds with SILKY white hairs
- season—maturing early summer

RIVER-LAKE MARGIN COMMUNITY

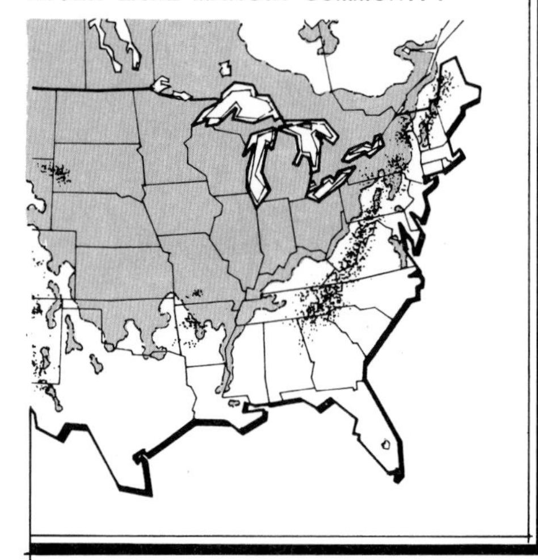

The Farmstead-Windbreak group is the only community of plants listed in this book that does not occur naturally.

Some would argue that it does not belong in the same category with native plant communities; that may be true. However, those of us who live in the Midwest have come to think of evergreens and other windbreak species as natural to our environment, because they protect us so well from the ever-present wind.

Much of the Midwest consists of gently rolling farmland with few trees to block the prevailing winter winds. Since these winds come from the northwest, windbreaks are planted on the north and west sides of farmsteads to break the force of the wind.

An ideal windbreak consists of three or more rows of plants spaced 20 feet apart, with 12 to 18 feet between rows. Taller trees are placed in the center rows. Evergreens are usually used here because they hold their needles throughout the winter. Some sites, however, are not suitable for evergreens. In areas with heavy soils or poor drainage, broadleaf trees can be used in the center rows instead. Small trees and shrubs are used in the outer rows and at the ends of windbreaks. They are used primarily to maintain the desired density of the windbreak close to the ground.

Shelterbelts were also created in the 1930s to to help reduce soil erosion by the wind. These wide belts of plants were placed at the edge of fields stretching for miles through North and South Dakota and Canada.

Farmstead-Windbreak Community

Trees: Deciduous (over 48′)

Trees: Understory (12–48′)

Trees: Evergreen

Northern Catalpa is one of the Midwest's most unusual native trees. It has the largest simple (one piece) leaf of any forest tree, sometimes growing as long as 1 foot. The leaves develop in late spring, usually not appearing until other trees are in full foliage. The flowers also bloom late and do not make an appearance until the warmer days of June.

As a boy in Kansas, I remember that the blooming of Catalpa marked the beginning of summer. The park across the street always opened its supervised play activities as the Catalpa was laying down its floral carpet along the paths by the tennis courts.

Catalpa's large white flowers grow in heavy clusters at the outer ends of the branches. This tree ranks at the top of the list of large canopy trees for the beauty of its bloom. The huge seedpods are almost as wonderful as the flowers; these long, slender cylinders dangle on the tree all winter and open in the spring to discharge many silvery winged seeds.

Northern Catalpa comes from a very narrow region. It grows from Indiana to Arkansas along the confluence of the Ohio and Mississippi rivers. It is a tall, upright tree of lowland terraces and alluvial flats. The reason it is found in almost every town and many farmsteads in the Midwest is primarily because its wood is resistant to decay. Because of this, Catalpas were planted as a wood supply for railroad ties.

Northern Catalpa grows fast but lives a short time, usually maturing in fifty to seventy-five years.

Catalpa speciosa · Northern Catalpa

habitat . . .
FOREST. lowland wet-mesic, alluvial bottomlands and farmstead windbreaks
- zone – 5a

form . . .
IRREGULAR. conical, large canopy tree (75–100′)
- branching – TWISTED, ascending limbs
- twig – stout, red-brown with small, swollen red-brown buds (inconspicuous), large leaf SCAR; conspicuous lenticels
- bark – smooth, BROWN with thin, irregular, scaly ridges; shallow-furrowed

foliage . . .
WHORLED. simple, CORDATE in 2s and 3s, yellow-green leaves (6–12″) with entire margins; coarse, tomentose beneath with thick, long petioles
- color (fall) YELLOW-green
- season – deciduous

flower . . .
CLUSTER. large, white (2 – 2¼″), BELL-shaped corolla (5–8″) with orange (2) stripes and purple spots
- sex – monoecious, FRAGRANT

fruit . . .
LEGUME. long, (8–20″), papery, tan-brown CIGAR-like pod with flat, brown seeds
- season – maturing autumn, PERSISTING through winter

FARMSTEAD-WINDBREAK

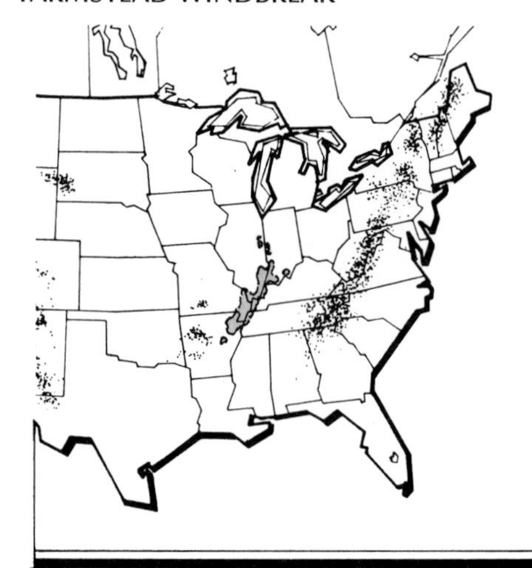

European Larch is a deciduous conifer, so it sheds its needles every autumn. This tree was imported from Europe and has been used in windbreaks and parks throughout the Midwest. Eastern Larch *(Larix laricina)* is a native tree, but it does not adapt as well to dry conditions and prefers bogs or wet soils in open, sunny meadows.

Larch needles are about an inch in length and appear to be whirled around small twig-like projections. These needles are bright green in spring, blue-green in summer, and golden tan in fall.

In spring, the tree blooms just before the needles appear. The male flowers are golden yellow and the female flowers are deep pink or red. Both are quite small and subtle and are barely noticed from a distance. The female flowers develop into small cones that persist all winter, shedding their seeds while remaining on the tree.

Larch seed is the preferred food of Purple Finches, and Spruce Grouse feed on its buds and needles. It is the larch that Aldo Leopold refers to as "tamarack" in his chapter, "Smoky Gold," in *Sand County Almanac* (1977).

European Larch grows rapidly and can survive for 200 years.

habitat . . .
FOREST. lowland wet, stream edge, swamp and lake margin
- zone—2

form . . .
CONICAL. irregular, small canopy tree (50–75′)
- branching—slender, horizontal, open pendent limbs
- twig—stout, yellow, gray-brown, smooth with short, lateral, SPUR bracts having small buds; red-brown
- bark—scaly, reddish brown with fine, flaky scale appearance

foliage . . .
NEEDLE. soft, slender, 3-ANGLED, light blue-green leaves (3/4–1″) appearing in WHORLS or clusters on spur-like twigs; glossy
- color (fall) YELLOW, amber-gold
- season—deciduous

flower . . .
CONE. (male) yellow, small; (female) purplish RED (3/8–1/2″), curved upright along twigs
- sex—monoecious

fruit . . .
CONE. ovoid, chestnut, tan-brown, ERECT (1/2–3/4″); stalkless with rounded cone scales; overlapping
- season—maturing autumn, dropping second year

FARMSTEAD-WINDBREAK

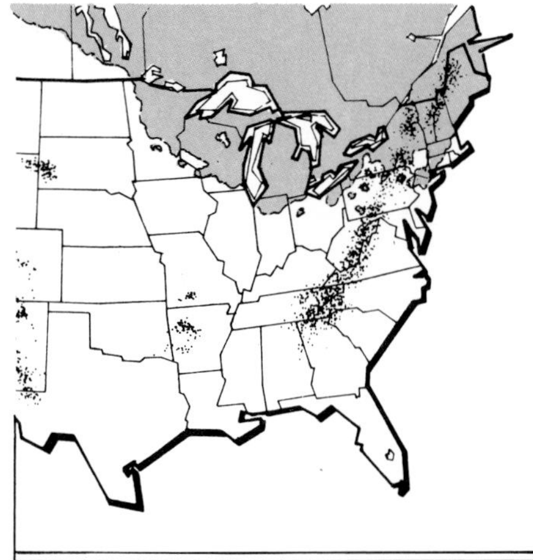

Larix decidua · European Larch

White Poplar was originally found in Turkestan and was widely used in Europe (Schoonover, 1951). It was introduced to this country by early New Englanders because it reminded them of home.

In the 1880s, White Poplar was one of the most common street trees, but today it is more of a curiosity. It lost favor with later generations in North America because its many suckers spring up in gardens and lawns. Even with constant pruning the suckers cannot be controlled, and many cities declared the plant a nuisance. Some even went so far as to prohibit its planting by ordinance.

White Poplar is, however, a beautiful, open-crowned tree that reaches heights of 100 feet. It has dark, shiny, green leaves above, and a velvety, silver-white wool beneath. As with most poplars, the leaves flutter in the slightest breeze. Another pleasing effect is created when sunlight reflects from the leaves much as it does on rippling water. The leaves are coarsely toothed, often lobed, and maple-like in appearance.

The bark of White Poplar is a second distinguishing mark. The branches and upper trunk are smooth and greenish-white. On mature trees, the bark on the lower trunk is thick and rigid. White Poplar is a fast-growing tree that can live well over 100 years.

Populus alba · White Poplar

habitat . . .
SAVANNA. upland dry, lowland wet-mesic, south and west facing slopes
- zone – 3b

form . . .
IRREGULAR. globular, large canopy tree (75–100′)
- branching – MASSIVE, irregular branches
- twig – smooth, yellow, green-gray, having diamond-shaped lenticels with sharp-pointed, red-brown buds; resinous
- bark – CHALKY-white with coarse-fissured ridges; furrowed base with black horizontal streaks

foliage . . .
ALTERNATE. simple, DELTOID, (3–5) lobed, dark green leaves (2½–4″); FLAT base with sharp-pointed margin; woolly, white, FELT-like beneath
- color (fall) YELLOW, wine-red
- season – deciduous

flower . . .
CATKIN. (male) crimson-red (1½–3″), hairy; (female) yellowish green (1½–2″), pendulous
- sex – dioecious, appearing before leaves

fruit . . .
CAPSULE. papery, tan (⅛–3⁄16″), EGG-shaped pods with white, COTTONY seeds
- season – maturing early summer

(Introduced from Europe and Asia)
FARMSTEAD-WINDBREAK

Black Locust is a small to medium-sized, columnar tree that is found in dry upland soils at the edges of woods and in open pastures. It has been used in windbreaks and woodlots and for erosion control throughout the Southeast, where it is native, and in the Midwest. It can now be found in all states east of the Rocky Mountains. Black Locust is a tree of contrasts. People either love it or hate it. Many love its unique appearance, its feathery foliage, and the intense fragrance of its white, pea-like blossoms. Others scorn it as a pest tree that forever sprouts from seeds and suckers where it isn't wanted. Black Locust also suffers from disfiguration by borers and bears sharp thorns along its branches and young trunk.

The freshly cut wood is rather soft, but after seasoning or curing it becomes exceedingly hard and resistant to the effects of exposure. Black Locust is an excellent wood for fuel and fence posts. It grows very quickly but is short-lived.

habitat . . .
FOREST, SAVANNA. upland dry, south and west facing slopes
•zone – 3b

form . . .
IRREGULAR. columnar, small canopy tree (50–75′)
•branching – OPEN, ascending limbs
•twig – stout, red-brown, slightly zig-zag, armed with paired SPINES (¼–½″) having minute rust-brown buds
•bark – FIBROUS, red-brown, black, deep-furrowed with rough scaly ridges; forking

foliage . . .
ALTERNATE. pinnately COMPOUND, leaflets (7–19) elliptic-oval, dark blue-green leaves (½–¾″) with entire margin; pale beneath, FOLDING at night
•color (fall) YELLOW-green
•season – deciduous

flower . . .
SPIKE. large, creamy white (¾″), PEA-like clusters in pyramidal (4–8″) spikes; DROOPING at base of leaves
•sex – dioecious, fragrant

fruit . . .
LEGUME. thin, tan-brown, FLAT (2–4″), papery pod; splitting open
•season – maturing in autumn

FARMSTEAD-WINDBREAK

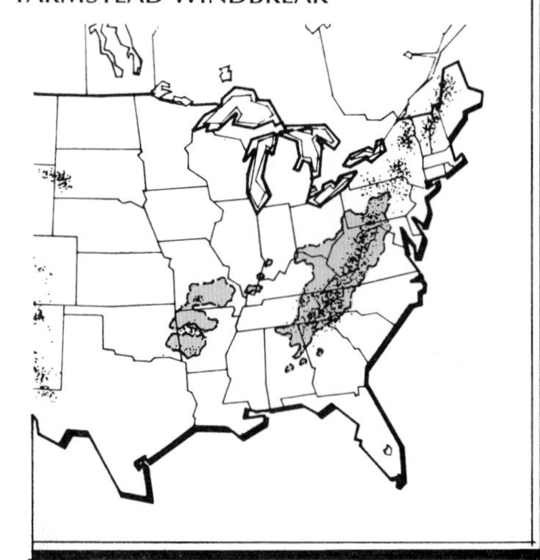

Robinia pseudoacacia · Black Locust

Russianolive is an ornamental that was introduced to this country from Europe and Asia during colonial times. This tree is believed to be the wild olive of classical literature (Keeler, 1903).

Russianolive was imported because of its ornamental features, such as its attractive silver-gray leaves, its silver olive-shaped berries, and its small yellow flowers that give off a rare fragrance. In fact, because of its fragrance, the Portuguese call it the Tree of Paradise. Because of its hardiness and resistance to drought, Russianolive has been used extensively in windbreaks and shelterbelts.

In its native region, Russianolive is a tree that attains heights of up to 50 feet. Here, it is both tree and shrub, seldom reaching more than 20 to 25 feet.

Russianolive is a fast-growing, short-lived tree. It is often used as an ornamental with an evergreen background or occasionally in combination with purple-leafed plum for contrast.

habitat . . .
FOREST, SAVANNA. upland dry, windbreak and fencerow planting
- zone—3a

form . . .
IRREGULAR. globular, small canopy tree (50–75′)
- branching—DENSE, ascending limbs
- twig—slender, silvery becoming REDDISH brown, ending in short, long-pointed spine
- bark—thin, silvery-gray to rust-brown, shredding in long LINEAR strips; trunk somewhat twisting

foliage . . .
ALTERNATE. simple, elliptic-LANCEOLATE gray-green leaves (1½–3½″) with entire margin; silvery, scaly beneath
- color (fall) SILVER-gray
- season—deciduous, PERSISTING through winter

flower . . .
CLUSTER. globular, silvery yellow (¼–⅜″), BELL-shaped clusters scattered along twig
- sex—monoecious, FRAGRANT

fruit . . .
BERRY. small, yellow-brown (⅜–½″), EGG-shaped with silvery bloom; base of leaf axils
- season—maturing late summer/autumn

(Introduced from Europe and Asia)
FARMSTEAD-WINDBREAK

Elaeagnus angustifolia · Russianolive

The **Osageorange** has many names and all are very descriptive of the plant. It was introduced into cultivation by some of the earliest settlers in St. Louis with specimens procured from the Osage Indians (Weed, 1936). The tree seems to have adopted orange or yellow as its favorite color, for it is yellow from root to fruit. The inner bark is dark orange, the sapwood is lemon-colored and the heartwood is a brilliant orange. The ripened fruit is yellowish green and the leaves in autumn turn a clear yellow. Even the roots, where exposed, have an orange bark. Indians dyed their blankets yellow using an extract made from its wood.

The French settlers of Louisiana gave the tree the name *Bois d'arc* or Bow Wood. The name was later changed to Bowdock but the meaning was the same, and the Indians used the tree to produce war clubs and bows. Even today its most valuable commercial use is in the manufacture of archery bows.

The name "Hedge-Apple" is descriptive of the tree's huge fruits and the use of the tree in the development of the great prairie region beyond the Mississippi. Because the forests of Ohio were disappearing, rail fencing was being discarded. Barbed wire was not yet invented, and the news of a thorny hedge substitute was gladly received. Farmers everywhere planted it, and for a generation or two it was the best fencing material throughout the Midwest. When wire fence came into use, the Osageorange lost its appeal but it had become naturalized in the landscape. Osageorange makes a good specimen tree: 50 feet tall, fast growing and long-lived, with the female trees producing the large, sticky fruit.

Maclura pomifera · Osageorange

habitat . . .
SAVANNA. upland dry, lowland wet-mesic, windbreak and fencerow planting
- zone – 5a

form . . .
GLOBULAR. irregular, large understory tree (35–50′)
- branching – ARCHING to recurving limbs
- twig – stout, orange-brown, ZIG-ZAG, armed with short-SPURRED (¼–1″) spine having small ball-shaped buds; brown
- bark – deeply furrowed, yellow, orange-brown with FIBROUS, cross-check ridges

foliage . . .
ALTERNATE. simple, oblong-ovate, GLOSSY, bright green leaves (2½–5″) with entire margin; thick, paler beneath, long-pointed tip
- color (fall) YELLOW
- season – deciduous

flower . . .
CLUSTER. small, yellow-green, (1″) dia., globular, dense clusters
- sex – dioecious

fruit . . .
MULTIPLE. large, yellow-green, (3–5″) dia., globular BALL covered by fibrous hairs; MILKY juice when cut
- season – maturing in autumn

FARMSTEAD-WINDBREAK

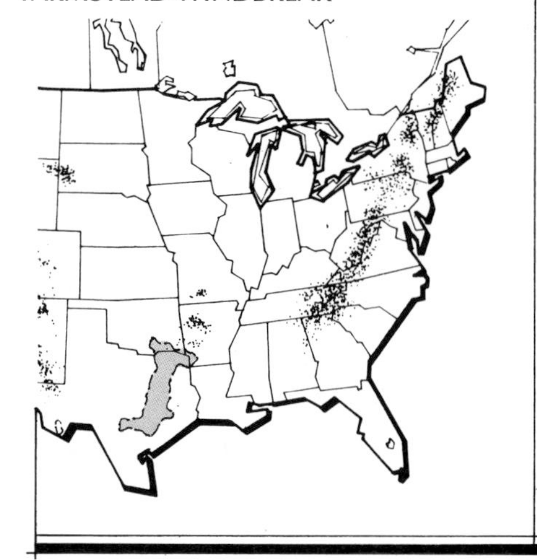

White Fir is an evergreen tree native to mountain ranges of the far West. There, in its natural habitat, it becomes a majestic tree of 100 to 180 feet tall and 3 to 5 feet thick at the base. Such growth by the White Fir is unequaled anywhere else in the world.

Young specimens are definitely pyramidal in character. With age, however, this characteristic gradually disappears and the crown develops a rounded form.

The tree has beautiful blue-green or whitish needles that curve gracefully upward. Another minor surprise is the rich red color of the pollen-producing male cones that ripen on the lower branches. The female cones stand upright on the branches and do not drop intact from the tree as do pines and spruce, but instead disintegrate on the tree, losing their scales and then persisting as ungainly spikes.

The White Fir is a tree that will adapt to dry or moist soils, but it requires less moisture than other firs. It is a rapid grower, long-lived and should be a first choice when consideration is being given to the selection of a specimen evergreen.

Abies concolor · White Fir

habitat . . .
FOREST. upland mesic-dry, dry rocky slopes and stream edge
- zone – 4a

form . . .
CONICAL. large canopy tree (75–100′)
- branching – stiff, HORIZONTAL branches often having two upper trunk leaders
- twig – stout, yellow-green with CIRCULAR leaf scar having clustered blunt, resinous buds; red-brown in GROUPS of 3
- bark – ash, gray-brown, shallow furrowed with heavily FLATTENED ridges

foliage . . .
NEEDLE. flat, linear, 2-RANKED blue-green leaves (1½–2½″) with whitish lines (top and bottom) CURVING upwards in U-shape along twig, blunt-pointed, yellow tip; winter discoloration
- color (fall) SILVERY blue-gray
- season – evergreen; AROMATIC, strong lemon smell when crushed

flower . . .
CONE. (male) yellow, small rounded; (female) YELLOW-green (¾–1″), erect
- sex – monoecious

fruit . . .
CONE. erect, PURPLE, olive-brown, cylindrical (3–5″) with central stalk of SHEDDING scales on uppermost twigs
- season – maturing late summer, falling apart upon ripening

FARMSTEAD-WINDBREAK

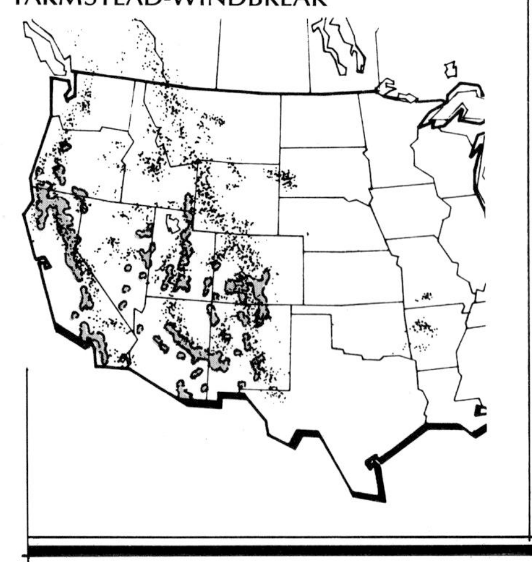

Norway Spruce is a common evergreen in northern Europe that grows in moist soils in cool, temperate regions. It has a distinct conical shape, like a church steeple. The lower branches, sweep close to the ground. The lesser branches dangle, giving the tree a weepy appearance. The needles are dark green, lustrous, sharp-pointed; and somewhat longer (by an inch) than those of native White, Red, and Black Spruce. The showy cones of Norway Spruce, the longest of the spruces at 4 to 7 inches in length, usually remain on the tree throughout the winter.

Norway Spruce grows rapidly and is free from insect and fungus problems. Seedlings are now being used for reforestation.

habitat . . .
FORESTS, SAVANNA. upland dry, lowland wet-mesic
- zone – 3a

form . . .
CONICAL. large canopy tree (75–100′)
- branching – dense, ASCENDING limbs
- twig – stout, gray-brown with rough PEG-like base; buds dark brown, rounded
- bark – dark gray-brown with IRREGULAR, shallow plates of thin, FLAKY scales

foliage . . .
NEEDLE. stiff, sharp-pointed, 4-sided, shiny dark green leaves (½–1″) with flattened, gracefully pendent leaflets; spreading on all sides
- color (fall) dark GREEN
- season – evergreen, AROMATIC

flower . . .
CONE. (male) yellowish brown (⅛–⅜″), clustered at twig tip; (female) pinkish MAROON, turning downward, green
- sex – monoecious

fruit . . .
CONE. long, brown, cylindrical (4–6″), woody; PENDULOUS
- season – maturing autumn, opening first year

(Introduced from Europe)
FARMSTEAD-WINDBREAK

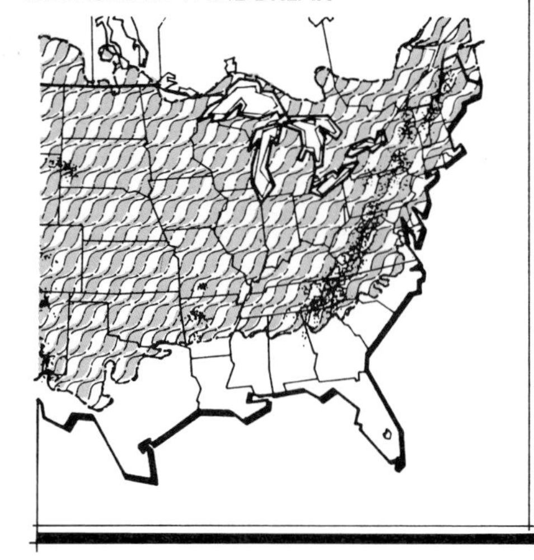

Picea abies · Norway Spruce

habitat . . .
FOREST. upland mesic, lowland wet-mesic and windbreak plantings
- zone – 2

form . . .
CONICAL. large canopy tree (75–100′)
- branching – sweeping, often DESCENDING limbs
- twig – slender, yellow-brown, rough PEG-like base with ovoid, chestnut brown buds
- bark – silver-gray; thin, IRREGULAR, scaly plates

foliage . . .
NEEDLE. single, stiff, sharp-pointed, 4-SIDED, blue-green leaves (¼–¾″) with whitish lines; curving up on sides of twig
- color (fall) bluish GREEN
- season – evergreen, AROMATIC

flower . . .
CONE. (male) RED, clustered (½–1″); (female) red-purple (½–¾″), ERECT, turning upward
- sex – monoecious

fruit . . .
CONE. small, tan-brown, CYLINDRICAL (1–2½″) with thin, flexible, smooth scale margin; seeds long-winged; paired
- season – falling at maturity

FARMSTEAD-WINDBREAK

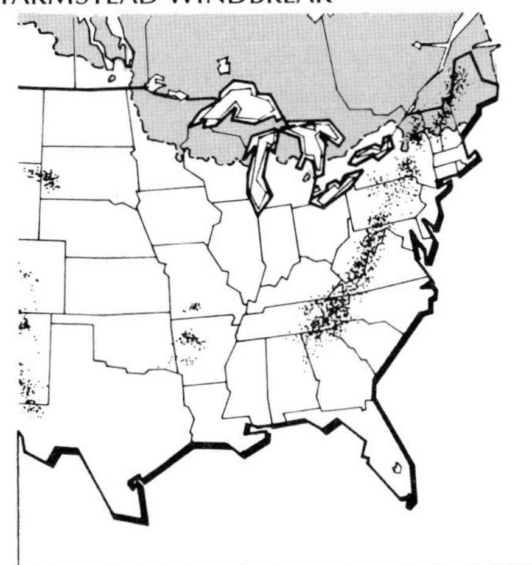

White Spruce is a northern tree typically found along stream borders and lake shores. It prefers moist sandy loams but grows in a variety of soils. White Spruce has a symmetrical, conical crown with many lower boughs reaching the ground.

The needles, which give the tree its name, are four-angled, short-pointed, and whitish; they grow in great abundance around the twig. The cones of White Spruce are 1 to 2 inches long, which is twice the length of Red and Black Spruce cones. These cones drop off after the seeds are shed and may be found beneath the tree in any season of the year.

Picea glauca · White Spruce

Colorado Spruce, native to the Rocky Mountain region, has adapted well to the lower elevations across the country. Colorado Spruce is a very dense tree that maintains a symmetrical, pyramid form with lower branches concealing the trunk.

The needles, at right angles from the twig, are extremely stiff, sharply pointed, and arranged in a spiral. (If needles are very blue, the plant is called Colorado Blue Spruce.) The cones are 2½ to 4 inches long, with paper scales having a wavy or irregularly toothed margin.

Colorado Spruce grows slowly for the first several years but later grows more rapidly, preferring full or partial sun. It is a widely used ornamental in the Midwest, as a lawn tree or in rows for screening and windbreaks.

habitat . . .
FOREST. upland mesic, mesic-dry, valley slopes and windbreak plantings
- zone – 2

form . . .
CONICAL. large canopy tree (75–100′)
- branching – sweeping, DESCENDING limbs
- twig – stout, orange, gray-brown with blunt, conical buds; yellow-brown
- bark – silvery gray-brown with broken, large, loose, papery scales

foliage . . .
NEEDLE. stiff, spirally arranged, 4-SIDED blue-green leaves (¾–1½″) at right angles to twig; slender, woody, PEG-like base
- color (fall) dark BLUE-green
- season – evergreen, AROMATIC

flowers . . .
CONE. (male) pale orange (⅝–1″), clustered; (female) purple-green (1–2″) with wavy-toothed scales
- sex – monoecious

fruit . . .
CONE. oblong, tan-brown, cylindrical (3–4″) with PAPERY scales of wavy margins; paired, long-winged seeds
- season – maturing autumn

FARMSTEAD-WINDBREAK

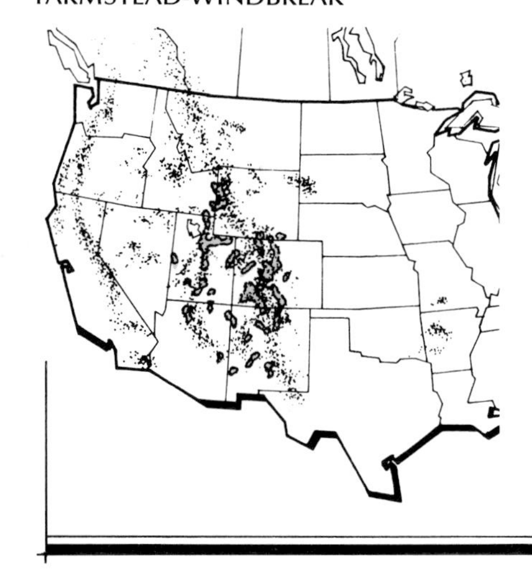

Picea pungens · Colorado Spruce

habitat . . .
FOREST. upland dry, steep rocky land and windbreak plantings
- zone – 2

form . . .
CONICAL. irregular, small canopy tree (50–75′)
- branching – open, RECURVING lower limbs
- twig – smooth, red-brown, FLEXIBLE with rounded (1/4″) red-brown buds; RESINOUS
- bark – irregular, brown-black, flaky PLATES with yellow-brown ridged patches

foliage . . .
NEEDLE. stiff, sharp-pointed clusters of 2, yellow-green leaves (3/4–1 1/2″) often TWISTED; basal sheath persistent
- color (fall) dark YELLOW-green
- season – evergreen

flower . . .
CONE. (male) reddish yellow (1/8″), clustered; (female) RED, small with prickly scales
- sex – monoecious

fruit . . .
CONE. tan-brown to black (1 1/2–2″); curving inward, often in pairs
- season – remaining closed many years

FARMSTEAD-WINDBREAK

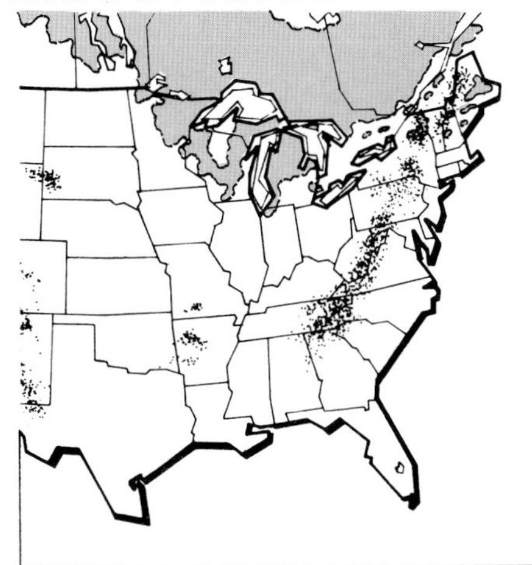

Jack Pine, the most Boreal of our native pines, is a small, irregular, conical evergreen seldom reaching 50 feet. Fast-growing but short-lived, reaching maturity in only sixty years, Jack Pine is often a scraggly outcast growing on barren ground, on rocky slopes, and in cool, boggy stretches where few plants can survive.

It has the shortest of pine needles, held in bundles of two that twist and separate in a flare. The small cones, seldom more than 2 inches, are curved, hug the twig, and are tightly closed.

Jack Pine requires fire to open its cones and release the seeds (Platt, 1968). Its cones, like those of Pitch Pine, hang on the branches for many years.

Pinus banksiana · Jack Pine

Lacebark Pine, a small, often multi-stemmed evergreen, is usually less than 30 feet tall. Introduced from China, it has been used for ornamental plantings and in urban windbreaks. A hardy tree, tolerating some shade and moist soils, it grows slowly.

Light green needles, in bundles of three, remain on the tree for five years, and give the branches a bottle-brush appearance. The cones have thick, spine-tipped scales that form a perfect rosette. Inside are large seeds, a winter food source for birds.

Its bark is multicolored (especially when wet), with splotches ranging from yellow-brown to gray-green. Peeling bark produces one color upon another.

habitat . . .
FOREST. upland mesic and mesic-dry
- zone – 5a

form . . .
CONICAL. bushy, small canopy tree (50–75′)
- branching – slender, horizontal, ASCENDING limbs
- twig – glabrous, bright gray-green with oval, sharp-POINTED scaly buds; tan-brown
- bark – smooth, lemon, yellow-green, EXFOLIATING large scales of CHALKY white patches; many trunked

foliage . . .
NEEDLE. stout, straight, shiny light yellow-green leaves (3–5″) often in clusters of 3; flexible
- color (fall) bright, dark GREEN
- season – evergreen, AROMATIC

flower . . .
CONE. (male) yellow (⅛–⅜″), clustered; (female) yellowish green
- sex – monoecious

fruit . . .
CONE. ovoid (2–3″), gray, yellow-brown, egg-shaped having scales with short prickles
- season – maturing second year

(Introduced from China)
FARMSTEAD-WINDBREAK

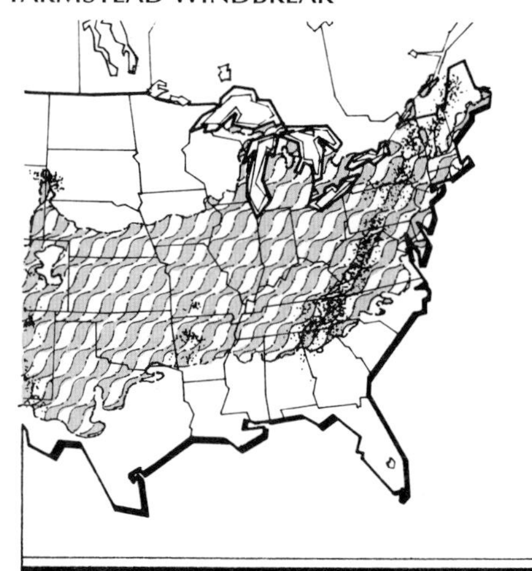

Pinus bungeana · Lacebark Pine

habitat . . .
FOREST. upland dry, steep rocky land
- zone—2

form . . .
OVOID. conical, small canopy tree (50–75′)
- branching—irregular, TAPERING limbs
- twig—stout, gray, flexible with oval-pointed buds of overlapping scales
- bark—smooth, silvery-gray with fine scaly plates

foliage . . .
NEEDLE. stiff, slightly CURVED, smooth, dark blue-green leaves (3/4–1 1/2″); clusters of 5
- color—dark BLUE-green
- season—evergreen, AROMATIC

flower . . .
CONE. (male) reddish yellow, small, cylindrical; (female) red (1/8–1/4″), clustered
- sex—monoecious

fruit . . .
CONE. oval, tan-brown, solitary (3–5″) or in groups of 2 or 3; scales with rigid, CURVED prickles
- season—opening at maturity

FARMSTEAD-WINDBREAK

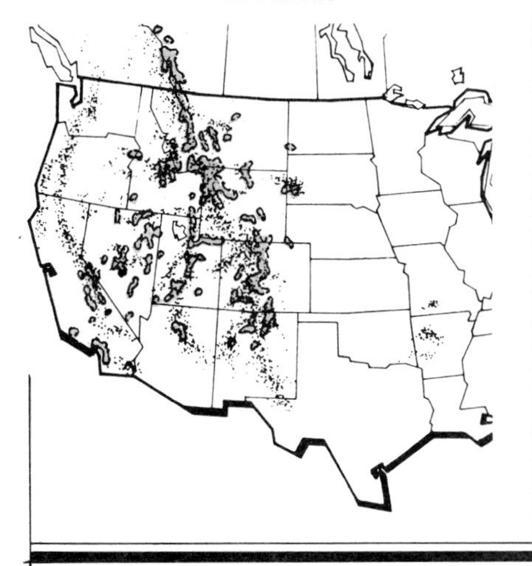

Limber Pine (Rocky Mountain White Pine) is a small evergreen tree, native to the dry upland slopes of the Rocky Mountain Forest region.

Limber Pine foliage is thick and beautiful, with needles in bundles of five, like those of Eastern White Pine. The cones are also similar, long and narrow, but the scales of Limber Pine cones are rounded and often reach 10 inches in length. The branches are extremely flexible and rubbery when young, giving reason for the common name, Limber, and the scientific name, *flexilis*.

Used in the Midwest for windbreaks and shelterbelts, this is a slow-growing tree that lives for 200 to 300 years.

Pinus flexilis · Limber Pine

Austrian Pine, one of the earliest foreign evergreens to be introduced in North America, has many advantages: it is hardy, grows rapidly, is easily obtained from nurseries, and tolerates polluted air quite well.

Austrian Pine needles are held in pairs. They are 4 inches long and have a shiny dark green surface. The cones are larger than Red Pine's, and the scales do not have the sharp-pointed spines that are common to the cones of Ponderosa Pine. The bark of Austrian Pine, its most recognizable feature, has a beautiful pattern of light gray to tan vertical plates.

Austrian Pine has been used for many years in protective windbreaks and ornamental plantings. It can be badly damaged by various needle diseases and is host to Pine Needle Scale.

habitat . . .
FOREST. upland mesic-dry and dry
- zone – 3a

form . . .
CONICAL. irregular, large canopy tree (75–100′)
- branching – HORIZONTAL, spreading to ascending limbs
- twig – stout, gray-brown with RESINOUS silvery white, long-pointed buds; overlapping scales
- bark – smooth, dark brown, shallow furrows with wide, flat SILVERY ridges

foliage . . .
NEEDLE. stiff, straight to slender, dark green leaves (3½–6″) often in clusters of 2; stout, not BREAKING cleanly
- color (fall) dark GREEN
- season – evergreen

flower . . .
CONE. (male) yellow-gold (¼–¾″) clusters; (female) RED (¼″) terminal pairs of 3s
- sex – monoecious

fruit . . .
CONE. yellow, tan-brown, conical (2–3″), EGG-shaped; scales ARMED with short spine
- season – opening and shedding after maturing

(Introduced from Europe)
FARMSTEAD-WINDBREAK

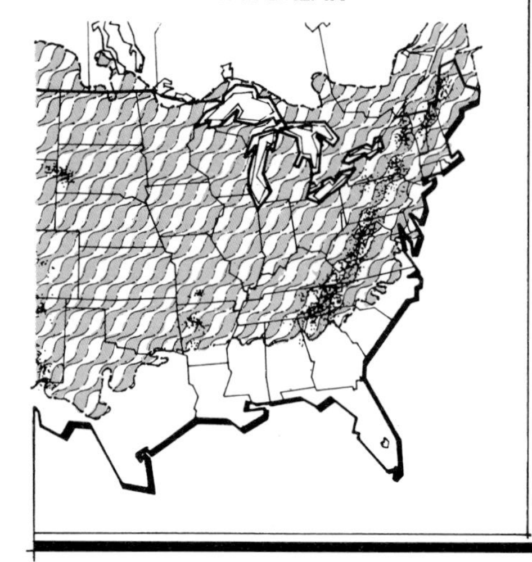

Pinus nigra · Austrian Pine

habitat . . .
FOREST. upland mesic-dry and dry, west and south slopes and windbreak plantings
•zone – 4b

form . . .
CONICAL. ovoid, large canopy tree (75–100′)
•branching – HORIZONTAL with recurving limbs
•twig – stout, orange-brown, turpentine ODOR when broken; (1″) RESINOUS buds
•bark – black, yellow-orange (old) with wide, shallow-furrowed, FLAKY plates

foliage . . .
NEEDLE. stout, FLEXIBLE, yellow-green leaves (5–10″) often in clusters of 2 or 3; forming TUFTS at ends of twigs
•color (fall) dark GREEN
•season – evergreen, AROMATIC

flower . . .
CONE. (male) reddish yellow, oblong (3/4–1″); (female) reddish purple (3/8″) terminal clusters
•sex – monoecious

fruit . . .
CONE. tan-brown, oval (3–6″) scales with tiny PRICKLES (keeled)
•season – open and shedding at maturity

FARMSTEAD-WINDBREAK

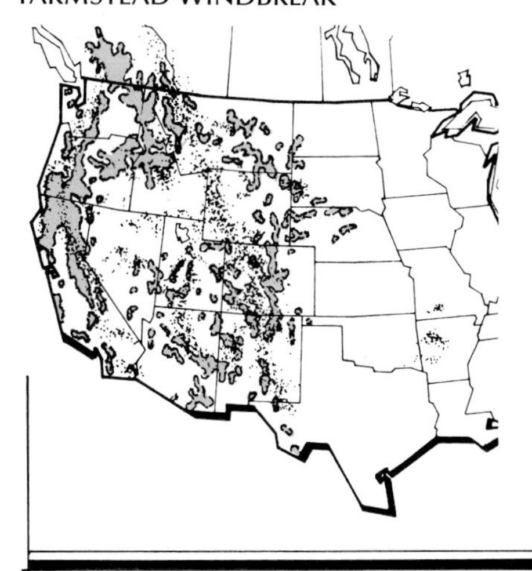

Ponderosa Pine, known by many names, including Western Yellow Pine, was discovered in 1804 by members of the Lewis and Clark expedition. David Douglas later gave it the name Ponderosa Pine (Collingwood and Brush, 1974).

In its natural range, this is a spire-shaped tree that grows 100 to 200 feet tall with a trunk 5 to 8 feet thick. In the Midwest, it grows 75 to 100 feet tall. The needles can be up to 10 inches long and are held in bundles of two or three. The small cones have sharp points on each scale. The bark is more interesting: No other pine has a patchwork of cinnamon-brown to yellow-orange scales that fit together like pieces of a puzzle.

Pinus ponderosa · Ponderosa Pine

Red Pine, or Norway Pine, is not from Europe. It may have earned the name Norway Pine from large stands found near the village of Norway, Maine.

A medium-sized tree that grows 60 to 80 feet tall, Red Pine is characterized by a long, straight, clean trunk and reddish-brown, flaky, scaly bark. The dark green needles, 4 to 6 inches long, are much longer than the needles of Scotch Pine. Unlike other pines, Red Pine needles snap cleanly when they are bent slowly, while others do not break. The cones are spineless, small, and compact.

Red Pine is resistant to disease. It grows rapidly, has a long life span, and reproduces easily from seed.

habitat . . .
FOREST. upland mesic-dry, steep rocky land and windbreak plantings
- zone–2

form . . .
CONICAL. ovoid, large canopy tree (75–100′)
- branching–horizontal, ASCENDING limbs
- twig–shiny, yellow-green with sharp-pointed buds; loose, overlapping HAIRY red-brown scales
- bark–gray, red-brown, wide, shallow-furrowed with long, FLAT, flaky ridges

foliage . . .
NEEDLE. straight, slender, bright green leaves (2–5″) set in clusters of 2; often clustered near branch ends
- color (fall) dark GREEN
- season–evergreen, AROMATIC

flower . . .
CONE. (male) purplish yellow, oblong (1–1¼″) clusters; (female) reddish purple (⅜″)
- sex–monoecious

fruit . . .
CONE. tan-brown, oval (2–2½″), often in pairs with UNARMED scales; often leaving basal scale
- season–opening and maturing autumn of second year

FARMSTEAD-WINDBREAK

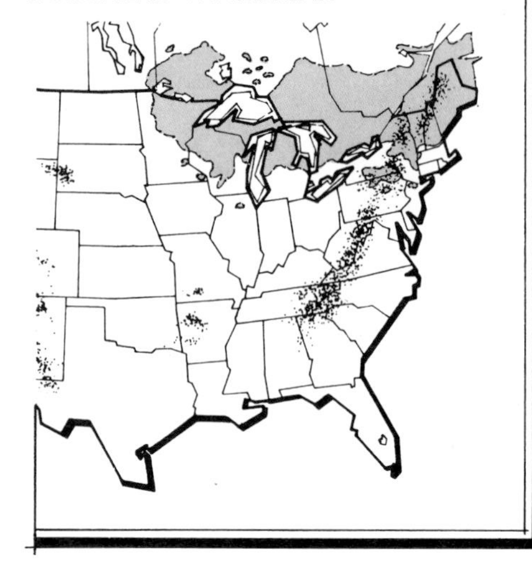

Pinus resinosa · Red Pine

habitat . . .
FOREST. upland mesic-dry, steep rocky land, west and south slopes and windbreak plantings
- zone – 5a

form . . .
IRREGULAR. globular, small canopy tree (50–75′)
- branching – drooping, TWISTING limbs
- twig – ridged, gray, brown-orange with clusters of sharp-pointed RESINOUS buds; scales chestnut brown
- bark – reddish brown, black, scaly to shallow-furrowed with wide, FLAT plates

foliage . . .
NEEDLE. stiff, often TWISTED yellow-green leaves (3–5″), bundles of 3s; often seen as TUFTS on twigs
- color (fall) dark YELLOW-green
- season – evergreen, AROMATIC

flower . . .
CONE. (male) red (3/8″) clusters; (female) pinkish-red (3/8″), often in whorls along twig
- sex – monoecious

fruit . . .
CONE. tan-brown, oval (2½–3″), PERSISTING in clusters; scales armed with RIGID prickle (keeled)
- season – opening at maturity, remaining attached

FARMSTEAD-WINDBREAK

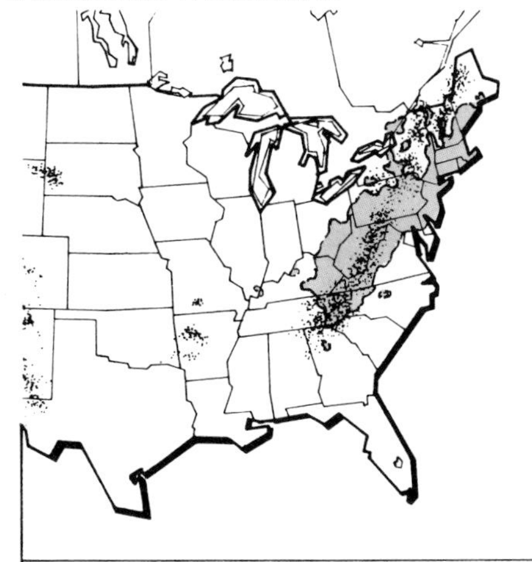

Pitch Pine, the only three-needled pine in the northeastern states, grows on dry sandy soils where other trees do not and attains heights of 40 to 70 feet. The cone has thick scales with prickly points. Cone production begins early in the tree's life. Cones are seldom dropped; instead they open and release their seeds in midwinter. The rough bark of Pitch Pine is broken into reddish-brown scales. It is so fire-resistant that it can be blackened and scarred for several feet above the ground with no apparent ill effects on the tree. Yet, when the bark is broken, great quantities of pitch flow. Pitch Pine grows slowly and lives 60 to 90 years, though some live for 200 years.

Pinus rigida · Pitch Pine

Scotch Pine, native to England and Scotland and also found in Europe and Asia, typically reaches heights of 60 to 90 feet and can grow 150 feet tall. When young, it is pyramidal, but with age it develops a broad, irregular, round-topped crown.

Its foliage, shorter than Austrian pine, is held in bundles of two. It has a looser appearance and grows more rapidly. Its orange bark, which is especially bright on the upper trunk and on the main branches, peels off in papery flakes.

The beautiful Scotch Pine, introduced in the Midwest for use in windbreaks, grows well in any type of soil, but is frequently bothered by Pine Needle Scale.

habitat . . .
FOREST. upland mesic-dry and dry, west and south slopes and windbreak plantings
- zone – 3a

form . . .
IRREGULAR. globular, small canopy tree (50–75′)
- branching – spreading, HORIZONTAL limbs
- twig – gray-brown, orange with oblong, RESINOUS small buds
- bark – thin, gray-brown to ORANGE with papery, EXFOLIATING limbs and upper trunk

foliage . . .
NEEDLE. stiff, slightly flattened, TWISTED yellow-green leaves (1½–3″) set in clusters of 2
- color (fall) YELLOW-green
- green – evergreen, AROMATIC

flower . . .
CONE. (male) yellow, round, globular clusters along shoots; (female) RED (⅛–¼″), terminal
- sex – monoecious

fruit . . .
CONE. yellow-brown, EGG-shaped (1½–2½″) with long, flat scales with RAISED point
- season – maturing second year

(Introduced from Europe and Asia)
FARMSTEAD-WINDBREAK

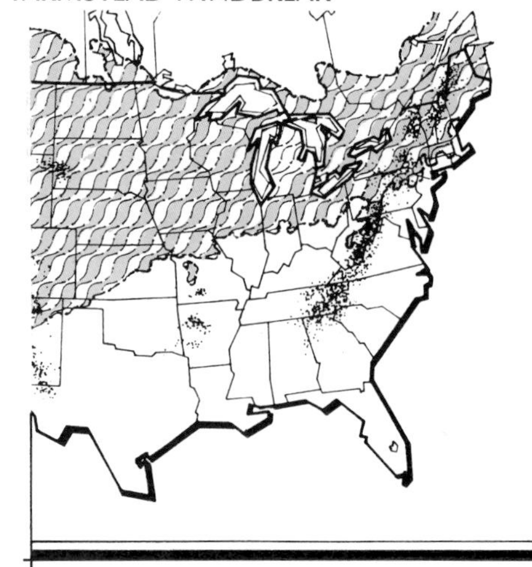

Pinus sylvestris · Scotch Pine

habitat . . .
FOREST. upland mesic, east and north facing slopes and moist, rocky land
- zone – 5b

form . . .
CONICAL. columnar, large canopy tree (75–100′)
- branching – straight, HORIZONTAL limbs
- twigs – slender, orange, gray-brown with long, sharp-pointed, PAPERY, scaled bud; red-brown
- bark – thick, reddish brown, deeply furrowed with SCALY ridges; smooth (young)

foliage . . .
NEEDLE. flexible, flattened, light green to bluish green leaves (3/4–1 1/4″) with rounded, blunt tip; spreading to spiral in 2-row arrangement
- color (fall) green, BLUE-GREEN
- season – evergreen, AROMATIC (fruity)

flower . . .
CONE. (male) yellow (3/8–1/2″), clustered along underside of shoots; (female) yellow–PINKISH green (3/4–1″), cylindrical with spreading scales at tip of old growth
- sex – monoecious

fruit . . .
CONE. cylindrical, tan-brown (2–3″), short-stalked, clustered PAIRS with rounded scales having 3-pointed, protruding BRACTS; seeds long-winged
- season – maturing autumn first year, PERSISTING through winter

FARMSTEAD-WINDBREAK

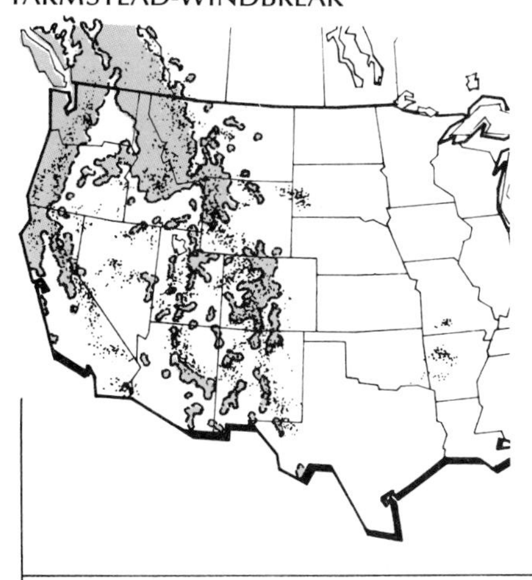

Common Douglasfir is also referred to as Red Fir, Oregon Pine, and Douglas Spruce. As its many names imply, there has been much argument about whether this tree is a fir, a pine, or a spruce.

For example, spruce needles are four-sided and stiff, and spruce cones hang down. Fir, on the other hand, has blunt, flat needles and cones that stand upright. However, Douglasfir needles are blunt and flat, but its cones hang down and have long, three-pointed seed-bracts that extend beyond the scales.

Common Douglasfir is a fast-growing evergreen tree, reaching heights of 80 feet in the Midwest and 180 feet in its native western mountain region. It ranks as one of America's tallest trees, next to redwoods and sequoias.

Pseudotsuga taxifolia · Common Douglasfir

Glossary

Growth Forms & habitat

Annual. Of only one season's duration.

Savanna

Association. A plant community of definite composition presenting a uniform appearance and growing in uniform habitat conditions.

Biennial. Maturing from seed in two growing seasons; usually blooming second year.

Boreal. Northern biogeographical region; from *boreas,* the North Wind.

Canopy. Shade-producing crown of trees; leaf and branch elements.

Overstory. Tree crown greater than 48 feet in height at maturity.

Understory. Tree crown greater than 12 feet, but not over 48 feet in height at maturity.

Community. An aggregation of living organisms having a mutual relationship among themselves and to their enviornment; a grouping of many individuals, usually of two or more species as in Oak-Hickory.

Cultivar. A cultivated variety or hybrid of a parent plant species, often having a slight difference as to fall color, flower color, or growth pattern.

Dioecious. Having male and female flowers on separate plants.

Ephemeral. Persisting for one day only; of short duration.

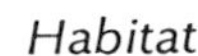

Habitat

Forb. Any herbaceous plant that is not grass or grass-like; annual, biennial or perennial.

Forest. Greater than 50 percent tree crown cover; successional or climax trees forming closed stands.

Genus. A group of related species, as the genus *Quercus* (oak) or the genus *Cornus* (dogwood); embracing respectively all the kinds of oak and all kinds of dogwood.

Grassland. Less than 5 percent tree crown cover; natural vegetation consisting largely of perennial grasses.

Habit. The general or physical character of a given species.

Habitat. The place or type of site where a plant or animal normally lives.

Mesic. Moist to moderate water content and well drained.

Xeric. Low water content; dry.

Herb. A non-woody plant that normally dies back to the ground at end of growing season; herbaceous.

Longevity. The length of time, with favorable condition, a plant can be expected to live.

Short-lived. Relatively short life; 100 years or less.

Medium-lived. 100 to 200 years.

Long-Lived. To endure for centuries, 200 years or more.

Monoecious. Having male and female flowers on the same plant.

Native. Originating naturally; plants that are indigenous to a particular region.

Naturalized. Spreading without cultivation to areas outside the native habitat.

Perennial. Producing flowers and seeds for more than two years; many surviving twenty years or more.

Prairie. An extensive level, slightly undulating, or rolling landscape, characterized by highly fertile soil and covered with coarse grasses and forbs.

Tallgrass. Grasses often soaring 8 feet in height with deep root systems; often found on the eastern prairies where rainfall is plentiful.

Midgrass. Grasses growing 2 to 4 feet in height; often found in bunches or clumps where rainfall is less abundant.

Shortgrass. Grasses growing only 16 inches high with shallow root systems; often found in regions where grasses compete well under dry conditions and under grazing.

Savanna. 5 to 50 percent tree crown cover; scattered trees, forest openings, and forest-prairie transition into either open plains or woodlands.

Species. A group of like individuals having common characteristics or qualities, as in Northern Pin Oak and Eastern White Pine.

Tundra. A treeless plain, characteristic of subarctic regions.

Vegetation. The plant cover of an area.

Zone. An area restricted by a range of annual average minimum temperatures; used in determining plant hardiness and adaptability.

community

association

Plant Form

Bunchgrass. Grasses growing in clusters with bare soil between groups of plants; often found in shortgrass prairies.

Columnar. Cylindrical, vertical axis much exceeding horizontal.

Conical. Approaching triangular in outline, broadest at base.

Creeping. Growing along the ground.

Erect. Upright habit of growth.

Globular. Rounded circular form, vertical and horizontal axis about equal.

Groundcover. Low-growing woody plants that trail on the ground.

Irregular. Asymmetrical, uneven outline.

Obovoid. Elliptic to egg-shaped, broadest at crown apex.

Ovoid. Elliptic to egg-shaped, broadest at base.

Prostrate. Lying on the ground, usually horizontal, flat; describing growing habit.

Twining. Winding spirally about a support.

irregular

globular

Branch Form

Ascending. Main primary branches diverging from the trunk at near 45-degree angles.

Axis. The main stem or line of development.

Crown. The upper mass or head of a tree, shrub, or vine; includes branches with foliage.

Descending. Main primary branches diverging from the trunk at angle below horizontal.

Downy. Having a coat of soft, fine hairs on buds, twigs, and leaves; pubescent.

Horizontal. Main primary branches predominately oriented at near 90-degree angles to the trunk.

Leader. The primary trunk of a tree.

Leggy. Multiple-stemmed plant devoid of foliage below; spreading by root propagation.

Modified twig. A stiff dwarf branch.

Pendulous. Having a drooping habit or appearance.

Picturesque. Naturally irregular or contorted.

Pith. The spongy or hollow center of the twig or stem; often chambered, diaphragmed, star or triangular shaped.

Spur. A short, projecting, sharp-pointed root, or branch of a twig.

Upright. Main primary branches stiffly vertical or diverging at slight angles.

Velvety. Having long, stiff, densely erect hairs; hairy.

Weeping. Secondary branchlets or tertiary branch tips and twigs pendent.

Winged. Having a thin expansion bordering or surrounding an organ.

Zig-Zag. Bent back and forth at nodes.

picturesque

pendulous

Bark type

Blocky. Having short, knobby, warty, flat-topped squarish plates or ridges.

Exfoliating. Thin or coarse strips or sheets peeling vertically or horizontally into irregular mottled patches.

Fibrous. Composed of or resembling fibers.

Furrowed. Grooved by shallow or deep longitudinal cracks becoming checked across into rectangular plate-like segments.

Lenticel. Having corky growths (vertical or horizontal) that admit air to interior of twig or branch.

Scaly. Regular, thin, papery flakes separated by a mesh pattern of fine, shallow, vertical, and horizontal fissures.

Smooth. Having a continuously even surface with a minimum of fissures.

Spine. Having sharp, mostly woody outgrowths in the position of a leaf or stipule.

Thorny. Having sharp unbranched or branched outgrowths in the position of lateral branches.

blocky

samara-key

lenticels

Bud type

Clustered. Crowded so as not to be clearly opposite, alternate, or whorled; often blunt, rounded, or sharp-pointed.

Domed. Having a large rounded or conical shape surrounded by leaf scar.

Lateral. Borne in the axil of the previous season's leaf.

Leaf Scar. A mark on a twig indicating the point where a leaf was once attached.

Naked. One without scales.

Papery. Resembling paper in thinness, usually dry or brittle, often large.

Pointed. Having a sharp tip.

Resinous. Secreting a sticky, gummy substance.

Scaly. Covered with or composed of many modified leaf scales or bracts.

Stalked. Positioned above short, slender, peg-like stem along twig.

Terminal. Occurring at the end, summit, or tip.

Woolly. Having long, flat, densely appressed hairs.

clustered

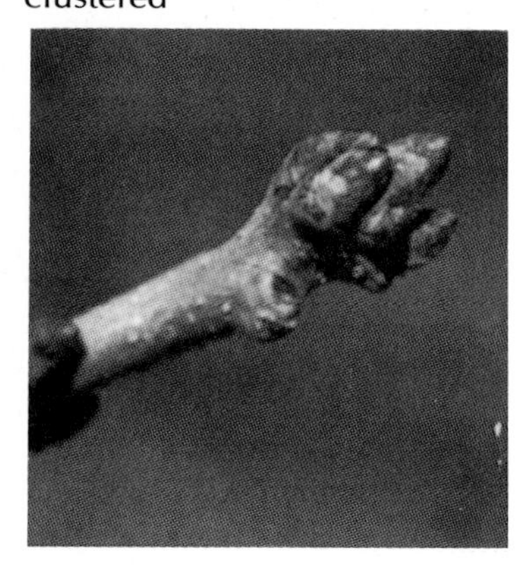

pointed

resinous

Root structure

Bulb. An enlarged, vertical underground shoot consisting mainly of fleshy leaf base or scales.

Corm. A solid, bulb-like, fleshy base of a stem.

Deep lateral. Deeply penetrating side-roots.

Knees. Spur-like rounded structures arising from roots of swamp-growing plants or trees; as in Common Baldcypress.

Legume. Plant members of the Pea Family having nodules on the roots where, in conjunction with rhizobium bacterium, they transfer nitrogen from the air into a form utilized by plants.

Rhizome. A thick, fleshy, horizontal, below ground (subterranean) stem; often specialized for food storage.

Shallow lateral. Side-roots forming fibrous material to 4 feet in depth.

Stolon. A horizontal stem, from the tip of which a plant will sprout; stoloniferous.

Taproot. A deep, carrot-shaped main root to 15 feet or more in depth.

Tendril. A slender, twining appendage enabling plants to climb.

Tuber. The thick, fleshy, underground portion of a stem; rhizome or stolon.

deep lateral

tendril

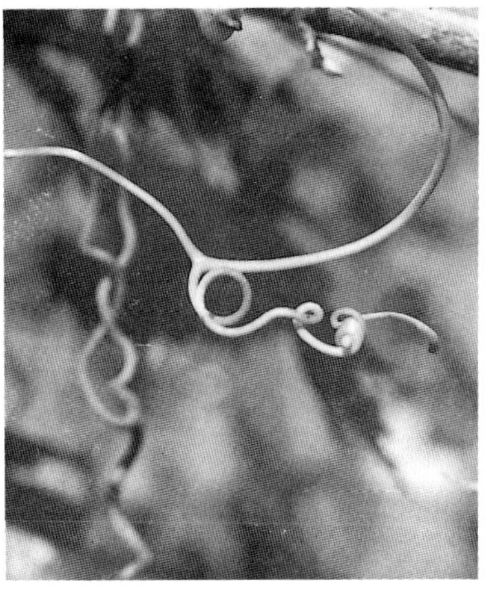

tuber

leaf type

Alternate. Occurring singly at different levels on the axis; not whorled or opposite.

Basal. Radiating arrangement of leaves close to the ground.

Blade. The elongated or expanded, flat-spreading portion of a grass leaf.

Compound. Composed of two or more blades; called leaflets.

Pinnately. With leaflets on opposite sides of a common axis or central stalk.

Palmately. With leaflets attached at the end of the petiole; radiating fan-like.

Trifoliate. With three leaflets.

Bipinnately. Leaf structure that is doubly or twice pinnate.

Fiddlehead. The curled young frond or leaf of a fern.

Fragrant. Agreeable or sweet scented; aromatic.

Frond. The leaf of a fern; often appearing compound or bipinnate (decompound).

Node. A joint on a stem represented by point of origin of leaf or bud; sometimes represented by a distinct leaf scar.

Opposite. Two leaves occurring at one node, on opposite sides of stem.

Perfoliate. Leaf appearing to be pierced by stem.

Persistent. Remaining attached after growing season.

Petiole. The stalk-like support of a leaf (blade).

Radial. Diverging in all directions from a central point.

Ranked. Foliage arranged in a flattened horizontal plane around a stem.

Sessile. Attached directly without petiole or peduncle.

Simple. Having one blade.

Stipule. Small appendages, often leaf-like, on either side of petiole base.

Whorled. Leaves occurring in a circle at a single node; three or more.

simple-alternate

trifoliate compound

pinnate compound

leaf shape

Acicular. Needle-like, slender, long, and pointed.

Cordate. Heart shaped.

Deltoid. Triangular, delta shaped.

Elliptic. Widest at middle, tapering to a point at both ends.

Keel. A sharp ridge or rib; resembling the prow of a boat.

Lanceolate. About four times as long as wide.

Linear. Narrow, sides about parallel.

Oblanceolate. About four times as long as wide, widest near tip end.

Oblong. Longer than wide, sides about parallel.

Obovate. Egg-shaped, widest near tip end.

Oval. Twice as long as broad, widest at middle with both ends rounded.

Ovate. Egg-shaped, widest near stem end.

Scaly. Small, short pointed leaves of some conifers, overlapping one another.

Spatulate. Rounded tip gradually narrowing to base.

cordate

obovate

acicular

leaf tip

Acuminate. Tapering to a long point.

Acute. Forming an angle of less than 90 degrees.

Cuspidate. With a firm, sharp point.

Obtuse. Forming an angle of more than 90 degrees.

Rounded. Without an angle or point.

rounded

leaf base

Acuminate. Tapering to a long point.

Acute. At a narrow angle.

Auricle. A small pointed or rounded projection of the sheath apex or base of the blade.

Clasping. Base wholly or partly surrounding the stem.

Cordate. Heart-shaped.

Glands. Minute globules, structures secreting resin or oily substances; often forming a small appendage or projection.

Ligule. A prominent claw-like projection or nest of hairs arising from the base of a leaf blade as it joins the sheath.

Oblique. With unequal sides (asymmetrical).

Rounded. Forming an ellipse or part of a circle.

Sheath. The tubular portion of the leaf that surrounds the stem; as in grasses.

cordate

clasping

leaf margin

Crenate. With rounded teeth.

Dentate. With teeth pointed outward.

Doubly Serrate. With small teeth on larger ones.

Entire. Without teeth or lobes.

Lobed. Deeply indented.

Revolute. Rolled toward the lower side.

Serrate. With teeth pointed forward; sawtoothed.

Sinuate. Strongly wavy.

Sinus. Space or recess between two lobes.

Undulate. Shallowly wavy.

lobed-sinuses

leaf surface

Glabrous. Smooth; without hairs.

Glandular. Bearing glands (appearing as dots).

Glaucous. Covered with waxy "bloom"; blue, white, or green.

Pubescent. Covered with short soft hairs.

Scabrous. Feels rough, like sandpaper.

Tomentose. With a dense woolly covering.

pubescent

leaf veination

Net-veined. Veins appearing on both sides of a main mid-rib, connecting with one another.

Palmate. Radiating fan-like from a common point.

Parallel. Running side by side parallel to leaf margin.

Pinnate. Forming an equal pattern along either side of mid-rib; feather-like.

parallel

Flower type

Axillary. Occurring at the junction of the leaves or branches.

Catkin. Flowers in the form of slender elongated cluster.

Clustered. Flowers occur in mostly small (small clustered) or mostly large (large clustered) clusters along the stem.

Cone. Small staminate and pistallate flowers clustered into tight cylindrical cones.

Cyme. More or less flat-topped cluster; central flowers blooming first.

Disk-flower. Small tubular flowers in the center part of a floral head; as in the Sunflower Family.

Flat-topped. Small flowers in rounded or flat-topped clusters mostly at end of twigs; corymb.

Floret. A small individual flower of a dense cluster, as in the grasses.

Fragrant. Agreeable or sweet scented; aromatic.

Imperfect. Lacking either stamens or pistil.

Inflorescence. Arrangement of flowers or flower clusters.

Panicle. An open, irregularly branched flower cluster on a central axis.

Perfect. A flower with both stamens and pistil.

Raceme. A long flower cluster on which individual flowers bloom on a small stalk.

Ray-flower. Bilaterally symmetrical flowers around the edge of the head in many members of the Sunflower Family; each ray-flower resembles a single petal.

Solitary. Flowers widely open with conspicuous, distinct petals.

Spadix. A dense spike of tiny flowers; fleshy column.

Spike. Flowers in elongated, pyramidal, or cylindrical, unbranched clusters; mostly at end of twigs.

Sterile. A flower incapable of producing seeds.

Terminal. Occurring at the tips of the twigs.

Trumpet. Bell-shaped and cup-like or tubular.

Umbel. Cluster of flower stalks occurring from the same point; like ribs of an umbrella.

cone

spike

flat-topped

Flower parts

Awn. A coarse, slender, bristle-like extension (barb).

Bract. A modified leaf; often situated at the base of a flower or inflorescence (involucre).

Bud. An unopened flower or an undeveloped leaf branch.

Culm. The flowering stem of a grass.

Keel. The lower two petals of flowers united; as in the Pea Family.

Pedicel. The stalk of a single flower in an inflorescence.

Petal. A division of the corolla; flat, usually broad, and brightly colored; showy.

Pistil. The central structure of the flower; the seed-bearing organ containing ovary, style, and stigma.

Pistillate. Provided with pistils but without stamens; composed of ovary, style, and stigma (female flowers).

Receptacle. The base of the flower where all flower parts are attached.

Scape. A leafless or naked flowering stem.

Spathe. A sheathing bract or pair of bracts around a spadix.

Spore. A reproductive body or detached cells of ferns, mosses, fungi, and similar plants; flowerless plant.

Staminate. Provided with stamens but without pistils; composed of anthers and filament (male flowers).

petal

spathe and scape

staminate-pistillate

Fruit type

pendulous

follicle

acorn

Achene. A small, dry, hard, non-splitting fruit with one seed.

Berry. Fruits with a soft, fleshy covering over the seed; pulpy indehiscent (non-opening).

Capsule. A fruit of more than one chamber, splitting lengthwise from one end along multiple seams.

Cone. A woody fruit having stiff, overlapping scales supporting naked seeds; primarily of coniferous plants.

Drupe. fleshy or pulpy fruit with a seed enclosed by a hard nut or stone, as in cherry or plum; a one-seeded, non-splitting fruit.

Follicle. An aggregate of small, fleshy pods on short, erect stems; resembling a cucumber.

Hip. The fleshy ripened receptacle of a rose, containing the bony achenes; often called rose-hips.

Key. A small, indehiscent, winged fruit; a samara.

Legume. An elongated bean-like pod, splitting open at maturity along two seams.

Multiple. A small, unwinged, sometimes plumed, one-celled, one-seeded fruit compounded to form a globular or ball-like head.

Nut or *Acorn*. Nut partially or wholly enclosed in a husk; papery, woody, leafy, or spiny in character.

Plume. A much-branched, silky, feather cluster of seeds.

Pod. A dry, dehiscent, linear, leguminous fruit.

Pome. A fleshy fruit, as in apple.

Samara. A small, indehiscent, winged fruit; a key.

Strobile. Slender, pendent or erect, catkin-like or cone-like fruits with papery, overlapping seeds.

Bibliography

Ahmadjian, Vernon.
Flowering Plants of Massachusetts. Amherst: Univ. of Massachusetts Press, 1979.

Aikman, J. M.
Native Shrubs and Vines of Iowa. Ames: Dept. of Botany, Iowa State Univ., 1957.

Summer and Winter Keys to Trees of the North Central States. Ames: Dept. of Botany, Iowa State Univ., 1957.

Ajilvsgi, Geyata.
Flowers of the Big Thicket: East Texas and West Louisiana. College Station: Texas A & M Univ., 1979.

Allen, Durward L.
The Life of Prairies and Plains. New York: McGraw-Hill, 1967.

Angier, Bradford.
Feasting Free on Wild Edibles. Harrisburg, Pa.: Stackpole Books, 1972.

Archer, S. G., and C. E. Bunch.
The American Grass Book: A Manual of Pasture and Range Practices. Norman: Univ. of Oklahoma Press, 1953.

Bailey, H. L.
The Standard Cyclopedia of Horticulture. 3 vols. New York: Macmillan Co., 1944.

Bare, Janet E.
Wildflowers and Weeds of Kansas. Lawrence: Regents Press of Kansas, 1979.

Barkley, T. M.
Atlas of the Flora of the Great Plains. Ames: Iowa State Univ. Press, 1977.

Borland, Hal.
A Countryman's Woods. New York: Alfred A. Knopf; 1983.

Braun, E. Lucy.
The Vascular Flora of Ohio. Columbus: Ohio State Univ. Press, 1967.

Brown, Clair A.
Wildflowers of Louisiana and Adjoining States. Baton Rouge: Louisiana State Univ. Press, 1972.

Brown, Lauren.
Weeds in Winter. Boston: Houghton Mifflin Co., 1977.

Budd, Archibald C., and Keith F. Best.
Wildplants of the Canadian Prairies. Ottawa: Canadian Dept. of Agriculture, 1964.

Cobb, Boughton.
A Field Guide to the Ferns: and Their Related Families. Boston: Houghton Mifflin Co., 1963.

Collingwood, G. H., and Warren D. Brush.
Knowing Your Trees. Washington, D.C.: American Forestry Assoc., 1974.

Core, Earl E.
Spring Flowers of West Virginia. Morgantown: West Virginia Univ. Press, 1981.

Courtenay, B., and J. H. Zimmerman.
Wildflowers and Weeds. New York: Van Nostrand Reinhold Co., 1972.

Cox, Donald D.
Common Flowering Plants of the Northeast. Albany: State Univ. of New York Press, 1985.

Curtis, John T.
The Vegetation of Wisconsin: An Ordination of Plant Communities. Madison: Univ. of Wisconsin Press, 1959.

Daubenmire, Rexford.
Plant Communities: A Textbook of Plant Syneocology. New York: Harper & Row, 1968.

Dean, Blanche, Amy Mason, and J. L. Thomas.
Wildflowers of Alabama and Adjoining States. University: Univ. of Alabama Press, 1973.

Dean, Charles C.
Flora of Indiana. Indianapolis: Dept. of Conservation, Div. of Forestry, 1940.

DeGraaf, Richard M., and Gretchin M. Whitman.
Trees, Shrubs and Vines for Attracting Birds: A Manual for the Northeast. Amherst: Univ. of Massachusetts Press, 1979.

Dorn, Robert D.
Manual of the Vascular Plants of Wyoming. New York: Garland Publishing, Inc., 1977.

Duncan, Patricia D.
Tallgrass Prairie: The Inland Sea. Kansas City, Mo.: Lowell Press, 1979.

Duncan, William H., and Leonard E. Foote.
Wildflowers of the Southeastern United States. Athens: Univ. of Georgia Press, 1975.

Dyas, Robert W.
The Spirit of the Savanna. Ames: Iowa State Univ., 1972. Traveling Photographic Exhibit.

Elias, Thomas S.
The Complete Trees of North America: Field Guide and Natural History. New York: Van Nostrand Reinhold Co., 1980.

Fassett, Norman C.
Spring Flora of Wisconsin. Madison: Univ. of Wisconsin Press, 1976.

Gilkey, Helen M., and La Rue J. Dennis.
Handbook of Northwestern Plants. Corvallis: Oregon State Univ. Bookstores, 1967.

Gleason, Henry A.
Illustrated Flora of Northeastern United States and Adjacent Canada. 3 vols. New York: Hafner Press, 1952.

Grimm, William C.
The Book of Shrubs. Harrisburg, Pa.: Stackpole Books, 1957.

The Illustrated Book of Trees. Harrisburg, Pa.: Stackpole Books, 1983.

Hartley, Thomas G.
The Flora of the "Driftless Area". 2 vols. Ann Arbor: Univ. Microfilms, 1976.

Harvill, A. M. Jr.
Spring Flora of Virginia. Parsons, W.Va.: McClain Printing Co., 1970.

Hersey, Jean.
The Woman's Day Book of Wildflowers. New York: Simon and Schuster, 1976.

Hightshoe, Gary L.
Forest Communities of Iowa: A Resource in Trouble. Ames: Iowa State Univ., 1980. Iowa Heritage Commission Traveling Photographic Exhibit.

Native Trees, Shrubs and Vines for Urban and Rural America. New York: Van Nostrand Reinhold Co., 1988.

Hosie, R. C.
Native Trees of Canada, Queens' Printer of Canada. 7th ed. Ottawa: Canadian Forestry Service, Dept. of Fisheries and Forestry, 1969.

Jones, Fred B.
Flora of the Texas Coastal Bend. Corpus Christi: Mission Press, 1975.

Keeler, Harriet L.
Our Northern Shrubs: How to Identify Them. New York: Charles Scribner's Sons, 1903.

Kelsey, H. P., and W. A. Dayton.
Standardized Plant Names. 2d ed. Harrisburg, Pa: J. Horace McFarland Co., 1942.

Kieran, John.
An Introduction to Trees. Garden City, N.Y.: Hanover House, 1954.

An Introduction to Wild Flowers. Garden City, N.Y.: Doubleday & Co., 1965.

Knobel, Edward.
Field Guide to the Grasses, Sedges, and Rushes of the United States. Mineola, N.Y.: Dover Pub., 1977.

Küchler, August W.
Potential Natural Vegetation of the Conterminous United States. New York: Amer. Geog. Soc., 1964.

Lakela, Olga.
A Flora of Northwestern Minnesota. Minneapolis: Univ. of Minnesota Press, 1965.

Lemmon, Robert S.
The Best Loved Trees of America. Garden City, N.Y.: Doubleday & Co., 1952.

Leopold, Aldo.
A Sand County Almanac. Madison: Tamarack Press, 1977.

Lerner, Carol.
Seaons of the Tallgrass Prairie. New York: William Morrow and Co., 1980.

Little, Elbert L.
The Audubon Society Field Guide to North American Trees (Eastern Region). New York: Alfred A. Knopf, 1979.

Lommesson, Robert C.
Nebraska Wildflowers. Lincoln: Univ. of Nebraska Press, 1973.

Miller, Howard A.
How to Know the Trees. Dubuque, Iowa: Wm. C. Brown Co., 1978.

Mohr, Charles.
Plant Life of Alabama. Montgomery: Brown Printing Co., 1901.

Morely, Thomas.
Spring Flora of Minnesota. Minneapolis: Univ. of Minnesota Press, 1969.

Moyle, J. B., and Evelyn W. Moyle.
Northland Wildflowers: A Guide for the Minnesota Region. Minneapolis: Univ. of Minnesota Press, 1977.

Nicholas, Stan, and Lynn Entine.
Prairie Primer. Madison: Univ. of Wisconsin Extension, Agricultural Bulletin, 1978.

Niering, Willie A., and Nancy C. Olmstead.
The Audubon Society Field Guide to Northern American Wildflowers (Eastern Region). New York: Alfred A. Knopf, 1979.

Odenwald, N. C., and J. R. Turner.
Plants for Designers: A Handbook for Plants of the South. 3d ed. Baton Rouge: Claitor's Publishing Division, 1987.

Oosting, Henry J.
The Study of Plant Communities. San Francisco. W. H. Freeman and Co., 1956.

Owensby, Clenton E.
Kansas Prairie Wildflowers. Ames: Iowa State Univ. Press, 1980.

Peterson, R. T., and M. McKenny.
A Field Guide to Trees and Shrubs. Boston: Houghton Mifflin Co., 1959.

Phillips Petroleum Company.
Pasture and Range Plants. Booklet Series (out of print). Bartlesville, Okla., 1957–1960.

Phillips, Roger.
Trees of North America and Europe. New York: Random House, 1978.

Platt, Rutherford.
Discover American Trees. New York: Dodd, Mead & Co., 1968.

Pohl, Richard W.
How to Know the Grasses. Dubuque, Iowa: Wm. C. Brown Co., 1978.

Porsild, A. E., and W. J. Cody.
Vascular Plants of Continental Northwest Territories, Canada. Ottawa: National Museums of Canada, 1980.

Preston, Richard J. Jr.
North American Trees. 4th ed. Ames: Iowa State Univ. Press, 1989.

Prior, Jean Cutler.
A Regional Guide to Iowa Landforms. Iowa City: Iowa Geological Survey, 1976.

Rehder, Alfred.
Manual of Cultivated Trees and Shrubs. 2d ed. New York: Macmillan Co., 1940.

Rogers, Julia E.
The Tree Book. Garden City, N.Y.: Doubleday, Doran & Co., 1931.

Roland, A. E., and E. C. Smith.
The Flora of Nova Scotia. Vol. 26. Halifax: Nova Scotia Museum, 1969.

Runkel, Sylvan T., and Alvin F. Bull.
Wildflowers of Iowa Woodlands. 1979. Reprint. Ames: Iowa State Univ. Press, 1987.

St. John, Harold.
Flora of Southeastern Washington and Adjacent Idaho. Anacortes, Wash.: Outdoor Pictures, 1963.

Sargent, Charles S.
Manual of Trees of North America. Vols. 1 and 2. Mineola, N.Y.: Dover Pub., 1961.

Schoonover, Shelley E.
American Woods. Santa Monica, Calif.: Walting and Co., 1951.

Simonds, R. L., and Henrietta H. Tweedie.
Wildflowers of Michigan. Michigan: 1978.

Smith, Arlo J.
Guide to Wildflowers of the Mid-South. Memphis, Tenn.: Memphis State Univ. Press, 1979.

Smith, Helen V.
Michigan Wildflowers. Bloomfield Hills, Mich.: Cranbrook Institute of Science, 1961.

Stephens, H. A.
Trees, Shrubs and Woody Vines in Kansas. Lawrence: Univ. of Kansas Pub., 1969.

Sternen, Thomas R., and Stanley W. Myers.
Oklahoma Flora. Oklahoma City: Harlow Publishing Co., 1937.

Steyermark, Julian A.
Flora of Missouri. Ames: Iowa State Univ. Press, 1965.

Spring Flora of Missouri. St. Louis: Missouri Botanical Garden, 1940.

Stubbendieck, J., S. L. Hatch, and Kathie J. Kjar.
North American Range Plants. Lincoln: Univ. of Nebraska Press, 1982.

Stupa, Arthur.
Wildflowers in Color: Eastern and Central America. New York: Harper & Row, 1965.

Van Bruggen, Theodore.
Vascular Plants of South Dakota. Ames: Iowa State Univ. Press, 1976.

Vance, F. R., J. R. Jowsey, and J. S. McLean.
Wildflowers across the Prairies. Saskatoon: Western Producers Prairie Books, 1977.

Viertel, Arthur T.
Trees, Shrubs, and Vines. Syracuse, N.Y.: Syracuse Univ. Press, 1970.

Voss, Edward G.
Michigan Flora: A Guide to the Identification and Occurrence—Part I (gymnosperms and monocotes). Bloomfield Hills, Mich.: Cranbrook Institute of Science, 1972.

Weaver, J. E.
Native Vegetation of Nebraska. Lincoln: Univ. of Nebraska Press, 1965.

Weber, W. A.
Rocky Mountain Flora. Boulder: Colorado Assoc. Univ. Press, 1976.

Weed, Clarence M., and Arthur L. Emmerson.
Our Trees: How to Know Them. Philadelphia: J. B. Lippincott Co., 1936.

Werthner, William B.
Some American Trees. New York: Macmillan Co., 1935.

Whaton, Mary E., and Roger W. Barbour.
A Guide to the Wildflowers and Ferns of Kentucky. Lexington: Univ. Press of Kentucky, 1971.

Whery, Edger T., and J. M. Fogg.
Atlas of the Flora of Pennsylvania: Philadelphia: Morris Arboretum, Univ. of Pennsylvania, 1979.

Wilkinson, R. E., and H. E. Jaques.
How to Know the Weeds. Dubuque, Iowa: Wm. C. Brown., 1979.

Woodward, Carol H., and H. William Rickett.
Common Wildflowers of the Northeastern United States. Woodbury, N.Y.: Barron's Educational Series, 1979.

Wyman, Donald.
Trees for American Gardens. New York: Macmillan Co., 1965.

Index

William C. Boon is Associate Professor of Landscape Architecture, Iowa State University. A native of Kansas, he graduated from Kansas State University with a bachelor's degree in landscape architecture and a second degree in technical agronomy. He holds a third degree, an M.L.A. from Iowa State University.

harlen d. Groe received his B.S. in landscape architecture at Iowa State University in 1978. He is a contributor to Gary Hightshoe's *Native Trees for Urban and Rural America* (1978) and several other publications. He is also a landscape designer, with a special interest in historic preservation studies and landmark documentation.

Notes :